Lecture Notes in Physics

For information about Vols. 1–131, please contact your bookseller or Springer-Verlag.

Vol. 132: Systems Far from Equilibrium. Proceedings, 1980. Edited by L. Garrido. XV, 403 pages. 1980.

Vol. 133: Narrow Gap Semiconductors Physics and Applications. Proceedings, 1979. Edited by W. Zawadzki. X, 572 pages. 1980.

Vol. 134: γγ Collisions. Proceedings, 1980. Edited by G. Cochard and P. Kessler. XIII, 400 pages. 1980.

Vol. 135: Group Theoretical Methods in Physics. Proceedings, 1980. Edited by K. B. Wolf. XXVI, 629 pages. 1980.

Vol. 136: The Role of Coherent Structures in Modelling Turbulence and Mixing. Proceedings 1980. Edited by J. Jimenez. XIII, 393 pages. 1981.

Vol. 137: From Collective States to Quarks in Nuclei. Edited by H. Arenhövel and A. M. Saruis. VII, 414 pages. 1981.

Vol. 138: The Many-Body Problem. Proceedings 1980. Edited by R. Guardiola and J. Ros. V, 374 pages. 1981.

Vol. 139: H. D. Doebner, Differential Geometric Methods in Mathematical Physics. Proceedings 1981. VII, 329 pages. 1981.

Vol. 140: P. Kramer, M. Saraceno, Geometry of the Time-Dependent Variational Principle in Quantum Mechanics. IV, 98 pages. 1981.

Vol. 141: Seventh International Conference on Numerical Methods in Fluid Dynamics. Proceedings. Edited by W. C. Reynolds and R. W. MacCormack. VIII, 485 pages. 1981.

Vol. 142: Recent Progress in Many-Body Theories. Proceedings. Edited by J. G. Zabolitzky, M. de Llano, M. Fortes and J. W. Clark. VIII, 479 pages. 1981.

Vol. 143: Present Status and Aims of Quantum Electrodynamics. Proceedings, 1980. Edited by G. Gräff, E. Klempt and G. Werth. VI, 302 pages. 1981.

Vol. 144: Topics in Nuclear Physics I. A Comprehensive Review of Recent Developments. Edited by T.T.S. Kuo and S.S.M. Wong. XX, 567 pages. 1981.

Vol. 145: Topics in Nuclear Physics II. A Comprehensive Review of Recent Developments. Proceedings 1980/81. Edited by T. T. S. Kuo and S. S. M. Wong. VIII, 571-1.082 pages. 1981.

Vol. 146: B. J. West, On the Simpler Aspects of Nonlinear Fluctuating. Deep Gravity Waves. VI, 341 pages. 1981.

Vol. 147: J. Messer, Temperature Dependent Thomas-Fermi Theory. IX, 131 pages. 1981.

Vol. 148: Advances in Fluid Mechanics. Proceedings, 1980. Edited by E. Krause. VII, 361 pages. 1981.

Vol. 149: Disordered Systems and Localization. Proceedings, 1981. Edited by C. Castellani, C. Castro, and L. Peliti. XII, 308 pages. 1981.

Vol. 150: N. Straumann, Allgemeine Relativitätstheorie und relativistische Astrophysik. VII, 418 Seiten. 1981.

Vol. 151: Integrable Quantum Field Theory. Proceedings, 1981. Edited by J. Hietarinta and C. Montonen. V, 251 pages. 1982.

Vol. 152: Physics of Narrow Gap Semiconductors. Proceedings, 1981. Edited by E. Gornik, H. Heinrich and L. Palmetshofer. XIII, 485 pages. 1982.

Vol. 153: Mathematical Problems in Theoretical Physics. Proceedings, 1981. Edited by R. Schrader, R. Seiler, and D.A. Uhlenbrock. XII, 429 pages. 1982.

Vol. 154: Macroscopic Properties of Disordered Media. Proceedings, 1981. Edited by R. Burridge, S. Childress, and G. Papanicolaou. VII, 307 pages. 1982.

Vol. 155: Quantum Optics. Proceedings, 1981. Edited by C.A. Engelbrecht. VIII, 329 pages. 1982.

Vol. 156: Resonances in Heavy Ion Reactions. Proceedings, 1981. Edited by K.A. Eberhard. XII, 448 pages. 1982.

Vol. 157: P. Niyogi, Integral Equation Method in Transonic Flow. XI, 189 pages. 1982.

Vol. 158: Dynamics of Nuclear Fission and Related Collective Phenomena. Proceedings, 1981. Edited by P. David, T. Mayer-Kuckuk, and A. van der Woude. X, 462 pages. 1982.

Vol. 159: E. Seiler, Gauge Theories as a Problem of Constructive Quantum Field Theory and Statistical Mechanics. V, 192 pages. 1982.

Vol. 160: Unified Theories of Elementary Particles. Critical Assessment and Prospects. Proceedings, 1981. Edited by P. Breitenlohner and H.P. Dürr. VI, 217 pages. 1982.

Vol. 161: Interacting Bosons in Nuclei. Proceedings, 1981. Edited by J.S. Dehesa, J.M.G. Gomez, and J. Ros. V, 209 pages. 1982.

Vol. 162: Relativistic Action at a Distance: Classical and Quantum Aspects. Proceedings, 1981. Edited by J. Llosa. X, 263 pages. 1982.

Vol. 163: J. S. Darrozes, C. Francois, Mécanique des Fluides Incompressibles. XIX, 459 pages. 1982.

Vol. 164: Stability of Thermodynamic Systems. Proceedings, 1981. Edited by J. Casas-Vázquez and G. Lebon. VII, 321 pages. 1982.

Vol. 165: N. Mukunda, H. van Dam, L.C. Biedenharn, Relativistic Models of Extended Hadrons Obeying a Mass-Spin Trajectory Constraint. Edited by A. Böhm and J.D. Dollard. VI, 163 pages. 1982.

Vol. 166: Computer Simulation of Solids. Edited by C.R.A. Catlow and W.C. Mackrodt. XII, 320 pages. 1982.

Vol. 167: G. Fieck, Symmetry of Polycentric Systems. VI, 137 pages, 1982.

Vol. 168: Heavy-Ion Collisions. Proceedings, 1982. Edited by G. Madurga and M. Lozano. VI, 429 pages. 1982.

Vol. 169: K. Sundermeyer, Constrained Dynamics. IV, 318 pages. 1982.

Vol. 170: Eighth International Conference on Numerical Methods in Fluid Dynamics. Proceedings, 1982. Edited by E. Krause. X, 569 pages. 1982.

Vol. 171: Time-Dependent Hartree-Fock and Beyond. Proceedings, 1982. Edited by K. Goeke and P.-G. Reinhard. VIII, 426 pages. 1982.

Vol. 172: Ionic Liquids, Molten Salts and Polyelectrolytes. Proceedings, 1982. Edited by K.-H. Bennemann, F. Brouers, and D. Quitmann. VII, 253 pages. 1982.

Lecture Notes in Physics

203

C. Marchioro
M. Pulvirenti

Vortex Methods in Two-Dimensional Fluid Dynamics

Springer-Verlag
Berlin Heidelberg GmbH 1984

Authors

C. Marchioro
Dipartimento di Matematica, Università di Trento, Povo
I-38050 Trento, Italy

M. Pulvirenti
Dipartimento di Matematica, Università di Roma "La Sapienza"
I-00185 Roma, Italy

Partially supported by M. P. I. and Italian C.N.R.

ISBN 978-3-540-13352-0 ISBN 978-3-540-38898-2 (eBook)
DOI 10.1007/978-3-540-38898-2

Originally published by Springer-Verlag Berlin Heidelberg New York in 1984

2153/3140-543210

INDEX

0. INTRODUCTION

These notes are based on lectures given at the meeting "Coulomb Systems and Fluid Dynamics" in Trento, May 27 - June 10, 1982 and concern the time evolution of planar incompressible fluids.

Our aim is to discuss some recent results on vortex methods, in view of the increasing interest, both theoretical and practical, in the argument.

The relevance of fluid mechanics in physics and applications is evident and we do not stress it here. Moreover the subject involves, quite naturally, several different mathematical techniques, such as the qualitative theory of ordinary differential equations, Hamiltonian systems, probability theory, stochastic processes and, obviously, partial differential equations, and stimulates analogies and connections with other fields of applied mathematics.

Our approach lies in the framework of mathematical physics: logical and mathematical rigour are taken into account, but we avoid straightforward, tedious or useless details.

Our analysis is mostly based on a well-known and old idea. A two dimensional incompressible ideal fluid behaves like a Hamiltonian system: each element of vorticity evolves under the action of a time-dependent vector field, solution of the Euler equations. It is then natural to consider the distribution of vorticity as formed by "particles", infinitesimal vortices, whose motion is decribed by Hamiltonian ordinary differential equations. We shall study such dynamical systems by which we construct solutions for the Euler equations.

When the vorticity is concentrated on points, the evolution of the fluid reduces to a Hamiltonian system with a finite number of degrees of freedom, called the vortex model. We investigate such a model and its relation to the Euler flow under suitable limits.

The vortex model is of interest not only theoretically: actually it provides a useful numerical algorithm for the evolution of ideal fluids.

This is, roughly, the content of Sections 1 - 5.

In the presence of viscosity a vortex dynamics can still be considered. In this case the vortex motion is described by stochastic differential equations. The stochasticity, due to the viscosity, diffuses and delocalizes the vortices. This point of view provides a good understanding of Navier-Stokes evolution in the absence of boundaries. (Sections 6 - 8).

The most physically interesting case, in which obstacles are present in the fluid, is more involved and has not been completely understood. It will be discussed in Section 9.

Although we use ideas from kinetic theory, our approach is completely macroscopic: a fluid is assumed to be a continuum and no critical discussion of this assumption will be made. The interesting problem of deducing the macroscopic properties of a fluid, starting from its microscopic structure, is still basically open, although rigorous results have recently been obtained for stochastic models [De.1].

Some parts of these notes (Sections 1 and 6) consist of classical and completely elementary arguments. We inserted them to make the exposition as self-contained as possible.

We do not pretend to give a complete review of the subject. For example, dynamics of vortex sheets, contour dynamics, statistical mechanical properties of systems of vortices, turbulent behaviour and other topics of great interest are only incidentally discussed or completely missing. We hope that the references quoted, although not at all complete, will lead the reader to a deeper knowledge of the subject.

It is a pleasure to thank G. Benfatto and R. Esposito for illuminating discussions, E. Omerti for the pictures of the vortex motion and G. Jona-Lasinio and F. Martinelli for having introduced us to large fluctuations theory.

April 1984

1. EULER EQUATIONS

In this Section we discuss the basic equations of Fluid Mechanics in a descriptive way and from a macroscopic point of view. The arguments we shall deal with are classical and may be found in any text book. We quote for example [Ba.1], [La.1] and, for a deeper mathematical level [Ch.1], [Hu.1], [Me.1], [Sh.1], [VM.1].

We assume a fluid confined in a region $D \subseteq \mathbb{R}^\nu$ ($\nu=2,3$ is the dimension of the physical space), described by the velocity field $\underline{u}(\underline{x},t)$, $\underline{x} \in D$, the density $\rho(\underline{x},t)$ and possibly other fields depending on the internal structure of the fluid, as temperature, thermodynamical pressure etc.

The laws of conservation of mass and balance of momentum give

$$\frac{d}{dt}\int_\Gamma \rho(\underline{x},t)d\underline{x} = -\int_{\partial\Gamma} \rho\underline{u}\cdot\underline{n}\, d\sigma \tag{1.1}$$

and

$$\int_\Gamma \rho(\underline{x},t)\underline{a}(\underline{x},t)d\underline{x} = \int_\Gamma \rho(\underline{x},t)\underline{F}(\underline{x},t)d\underline{x} + \int_{\partial\Gamma} \underline{\Phi}_{\underline{n}}(\underline{x},t)d\sigma \tag{1.2}$$

where

$\Gamma \subset D$ is an arbitrary smooth domain

$\underline{n}$ is the outward unit normal on $\partial\Gamma$

$d\sigma$ is the surface element

$\underline{a}$ is the molecular acceleration i.e.

$$\underline{a} \equiv \frac{D\underline{u}}{Dt} \quad \text{where} \quad \frac{D}{Dt} \equiv \frac{\partial}{\partial t} + \underline{u}\cdot\underline{\nabla}$$

$\underline{F}$ is a force for unit mass

(due to external forces and to the long range molecular interaction).

$\underline{\Phi}_{\underline{n}} d\sigma$ is the contact force (for example the forces arising from the interaction between the particles in the negative and positive side of $d\sigma$).

As consequence of Gauss-Green lemma we have from (1.1)

$$\frac{\partial}{\partial t}\rho(\underline{x},t) + \underline{\nabla}\cdot\rho\underline{u} = 0 \quad \text{(Continuity eq.)} \qquad (1.1)'$$

where $(\underline{\nabla})_i = \frac{\partial}{\partial x_i}$, $i = 1,\ldots,\nu$, is the divergence operator.

To apply the Gauss-Green lemma to Eq. (1.2),we need a representation theorem due to Cauchy:

$$\underline{\Phi}_{\underline{n}}(\underline{x}) = \sum_{i=1}^{\nu} \alpha_i \underline{\Phi}_{\underline{e}_i} \qquad (1.3)$$

where $\{\alpha_i\}$ are the projections of $\underline{n}$ on the basis $\{\underline{e}_i\}_{i=1}^{\nu}$.
From (1.2) and (1.3) we have

$$\rho(\underline{x},t)\frac{D\underline{u}}{Dt}(\underline{x},t) = (\rho\underline{F})(\underline{x},t) - \sum_{i=1}^{\nu} \frac{\partial}{\partial x_i} \underline{\Phi}_{\underline{e}_i}(\underline{x},t). \qquad (1.2)'$$

$\underline{\Phi}_{\underline{e}_i}$ is called stress tensor and contains almost all physical informations on the nature of the fluid.

For a class of fluids, called ideal, the stress tensor takes the form

$$\underline{\Phi}_{\underline{n}}(\underline{x},t) = p(\underline{x},t)\underline{n} \qquad (1.4)$$

where p is the pressure.

Hence, from (1.2)':

$$\rho(\underline{x},t)\frac{D\underline{u}}{Dt}(\underline{x},t) = (\rho\underline{F})(\underline{x},t) - \underline{\nabla}p \quad \text{(Euler eq.)} \tag{1.5}$$

The ideal fluids give a good description of real fluids in some circumstances.

Eq. (1.4) means that tangential frictions are neglected in the form of the stress tensor. A more accurate analysis, in the case in which the viscosity is not negligible, will be given in Section 6.

Obviously Eq. (1.1)' and (1.5) are not sufficient to determine the fields $\underline{u},\rho,p$. Another equation arising from thermodynamical considerations is introduced, to complete this system of equations, when $\underline{F}$ is known.

From now on we make the incompressibility assumption $\rho = \text{const.}$ and suppose $\underline{F}$ to be some known function describing external forces. Then:

$$\begin{cases} \dfrac{D\underline{u}}{Dt} = -\dfrac{1}{\rho}\underline{\nabla}p + \underline{F} & (1.6)_a \\[2ex] \underline{\nabla}\cdot\underline{u} = 0 & (1.6)_b \end{cases}$$

Eq.s (1.6) have to be complemented by boundary conditions. We assume

$$\underline{u}\cdot\underline{n} = 0 \quad \text{on} \quad \partial D \tag{1.7}$$

if D has boundary. This means that the fluid cannot penetrate the boundary. When D is not bounded, the asymptotic behavior of $\underline{u}(\underline{x})$ has to be specified when $|\underline{x}| \to \infty$.

We introduce now a basic concept for our analysis. We define

$$\underline{\xi} = \underline{\nabla} \times \underline{u} . \tag{1.8}$$

$\underline{\xi}$ is called vorticity and expresses how much the fluid is rotat ing. (For example for a rigid motion with angular velocity $\underline{\omega}$, $\underline{\xi} = 2\underline{\omega}$).

We suppose from now on, that $\underline{F}$ is a gradient field unless otherwise explicitely assumed. In this hypothesis we have two simple but useful theorems.

Theorem 1.1 (Kelvin). Let

$$I_{C_t}(t) = \oint_{C_t} \underline{u} \cdot \underline{ds} \qquad \text{(circulation)} \tag{1.9}$$

where C_t is a closed smooth curve, moving with the fluid. Then $I_{C_t}(t)$ is constant in time.

Proof. By direct computation

$$\frac{d}{dt} \oint_{C_t} \underline{u} \cdot \underline{ds} = \oint_{C_t} \frac{D\underline{u}}{Dt} \cdot \underline{ds} \tag{1.10}$$

and hence the thesis by the use of (1.6). □

A line which is tangent in each point to the vorticity is called a *vortex line*.

A *vortex tube* is a collection of vortex lines orthogonal to some closed curve.

Theorem 1.2 (Helmholtz). Let C_1 and C_2 by any two curves encircling the same vortex tube. Then

$$\oint_{C_1} \underline{u} \cdot \underline{ds} = \oint_{C_2} \underline{u} \cdot ds . \tag{1.11}$$

The above quantity is called strength of the vortex. (Constant in time by Kelvin's theorem).

Proof. (1.11) follows by Stokes theorem. □

The above theorems allows to visualize the time behaviour of the fluid: a growth of the velocity field is related to the stretching of a vortex tube.

Let us apply the operator curl = $\underline{\nabla}\times$ to both members of Euler Eq. $(1.6)_a$; using $(1.6)_b$ we obtain

$$\frac{D\underline{\xi}}{Dt} = (\underline{\xi}\cdot\underline{\nabla})\underline{u} \, . \tag{1.12}$$

A deep simplification arises when $\nu = 2$, or equivalently when $\nu = 3$ and the fluid has a plane symmetry and the velocity belongs to this plane. This case may occur approximatively in some physical situations (for example basins, oceans, atmosphere). For this reason and for technical simplicity, two dimensional fluids are widely investigated from numerical and theoretical points of view.

In this case only the z-component of $\underline{\xi}$ survives. Let us call it ω. Eq. (1.12) becomes

$$\frac{D\omega}{Dt} = 0 \, . \tag{1.13}$$

As we shall see, it is possible to establish a global existence and uniqueness theorem for the initial value problem associated to (1.13) in a completely satisfactory way, while in three dimensions, only short time existence and uniqueness theorems are available. In fact, the only Helmholtz theorem does not prevent the velocity to become singular.

From now on we shall deal with two dimensional incompressible fluids.

Let D be a simply connected domain. In virtue of $\underline{\nabla}\cdot\underline{u} = 0$ we can introduce a function ψ, called stream function, such that:

$$\underline{u} = \underline{\nabla}^{\perp}\psi \tag{1.14}$$

where $\underline{\nabla}^{\perp} = (\frac{\partial}{\partial x_2}, -\frac{\partial}{\partial x_1})$.

From (1.14) and definition of ω, we have

$$\Delta\psi = -\omega \tag{1.15}$$

The above Poisson equation allows us to reconstruct the velocity field knowing the vorticity ω, provided the boundary condition on ψ are specified. Let us suppose D bounded. By (1.7) we have that $\psi = \text{const.}$ on ∂D. (This constant will be assumed zero). Let $g_D(\underline{x},\underline{y})$ be the fundamental solution of the Poisson equation (i.e. $\Delta_{\underline{x}} g_D(\underline{x}-\underline{y}) = \delta(\underline{y})$, $g_D(\underline{x},\underline{y}) = 0$ if $\underline{x}$ or $\underline{y} \in \partial D$) then

$$\underline{u}(\underline{x},t) = \int_D (\underline{\nabla}_{\underline{x}}^{\perp} g_D)(\underline{x},\underline{y})\omega(\underline{y},t)d\underline{y}\,. \tag{1.16}$$

If $D = \mathbb{R}^2$ then

$$g_D(\underline{x},\underline{y}) \equiv g(\underline{x}-\underline{y}) = -\frac{1}{2\pi}\ln|\underline{x}-\underline{y}| \tag{1.17}$$

In this case the formula (1.16) gives only a part of the velocity field. It follows $\underline{u}(\underline{x},t) \to 0$ when $|\underline{x}| \to \infty$ if $\omega \in L_1 \cap L_\infty(\mathbb{R}^2)$. To take into account more general behaviors at infinity of the velocity field one has to add to (1.16) a potential flow

$\underline{v}(\underline{x}) \equiv (\nabla\varphi)(\underline{x})$, where φ is an harmonic function whose behavior at infinity is determined by the problem under consideration. The same is also true for general simply connected unbounded domains D. We notice that this potential flow describes actions localized at infinity.

Another possible interesting domain is the flat torus T^2. In this case the Green function exists and is

$$g_{T^2}(\underline{x},\underline{y}) = -\frac{1}{(2\pi)^2} \sum_{\substack{\underline{k}\in\mathbb{Z}^2 \\ \underline{k}\neq 0}} \frac{1}{k^2} \exp\{i2\pi\underline{k}\cdot(\underline{x}-\underline{y})\} . \tag{1.18}$$

We notice that g_{T^2} makes sense because $\int_{T^2} \omega d\underline{x} = 0$ by virtue of the circulation theorem. Also in this case the velocity field $\underline{u}$ is given by (1.16) modulo a constant overall flow.

Finally we remark that non simply connected domains may be treated in the same way with the aid of a multivalued stream function.

In what follows we shall consider fluids in simply connected domains and such that the potential part is absent, but our considerations can be generalized in a straightforward way.

Summarizing, our evolution problem takes the form

$$\begin{cases} \frac{\partial\omega}{\partial t}(\underline{x},t) + (\underline{u}\cdot\underline{\nabla})\omega(\underline{x},t) = 0 & (1.19)_a \\ \underline{u}(\underline{x},t) = \int (\underline{\nabla}^{\perp}_{\underline{x}} g_D)(\underline{x},\underline{y})\omega(\underline{y},t)dt & (1.19)_b \\ \omega(\underline{x},0) = \omega_0(\underline{x}) \qquad \underline{x} \in D, \quad t \geq 0. \end{cases}$$

We again point out that boundary conditions $\underline{u}\cdot\underline{n} = 0$ and

$\underline{u} \to 0$ if $|\underline{x}| \to +\infty$, are authomatically satisfied, under suitably assumptions on ω, because of the form of g_D.

The structure of the initial value problem (1.19), (called in the following Euler equations) suggests how to construct the solution. Given an arbitrary vector field $\underline{u}$, one solves the linear problem $(1.19)_a$. The solution of this problem determines, via $(1.19)_b$, a new velocity field $\underline{u}'$. The solution is therefore the fixed point of the map $\underline{u} \to \underline{u}'$ (see Section 3 below).

2. VORTEX MODEL

It is quite natural to study situations in which the vorticity profile is very sharply concentrated around some points called *vortices*. We schematize the situation by assuming

$$\omega(\underline{x})d\underline{x} = \sum_{i=1}^{n} a_i \delta_{\underline{x}_i}(d\underline{x}) \tag{2.1}$$

where $\delta_{\underline{x}_i}(d\underline{x})$ is the Dirac measure contered at point $\underline{x}_i \in D$, where D is an open region with smooth boundary, and a_i is a real number denoting the intensity of the vortex localized in $\underline{x}_i$. Then, the velocity field is

$$\underline{u}(\underline{x}) = \sum_{i=1}^{n} a_i \underline{\nabla}^{\perp} g_D(\underline{x},\underline{x}_i)\ . \tag{2.2}$$

$\underline{u}$ becomes singular when $\underline{x}$ approachs $\underline{x}_i$. In fact

$$g_D(\underline{x},\underline{y}) = -\frac{1}{2\pi}\ln|\underline{x}-\underline{y}| + \gamma_D(\underline{x},\underline{y}) \tag{2.3}$$

where γ_D is a smooth function in $D \times D$, and hence $\underline{u}$ diverges like $\frac{1}{|\underline{x}-\underline{x}_i|}$, [Co.1].

We notice that the vortex distribution (2.1) corresponds, in a three dimensional picture, to a distributions of vortex filaments parallel to the z-axis and with constant intensity.

We want to study the evolution of ω given by (2.1) by means of the Euler equations. Since ω is singular it is convenient to introduce a weak version of the initial value problem (1.19). Let f be a smooth test function, an application of the Gauss-Green lemma and $\underline{\nabla}\cdot\underline{u} = 0$ give

$$\begin{cases} \frac{\partial}{\partial t}\,\omega_t(f) = \omega_t(\underline{u}\cdot\underline{\nabla}f) & (2.4)_a \\ \underline{u}(\underline{x},t) = \int_D (\nabla^{\perp}_{\underline{x}} g_D)(\underline{x},\underline{y})\omega_t(d\underline{y}) & (2.4)_b \\ \omega_{t=0}(d\underline{x}) = \omega_o(d\underline{x}) \end{cases}$$

where

$$\omega_t(f) = \int_D \omega_t(d\underline{x})f(\underline{x}) \tag{2.5}$$

We see that the above initial value problem is meaningless, even is weak form. This because at time $t = 0$ the velocity field $\underline{u}$ does not make sense because of the singularity of the kernel in $(2.4)_b$.

We shall try to eliminate the infinite contribution due to the "self-energy" factor by constructing a new model (called vortex model) in which each vortex moves in the velocity field generated by all the others, neglecting in a suitable way the self interaction. We will justify this model by showing that it is in a sense equivalent to the Euler equation.

Consider one vortex in $\underline{x}_o \in D$. We approximate it by a sequence $\omega_n \to \delta_{\underline{x}_o}$ where ω_n are C^∞ spherically symmetric functions around $\underline{x}_o$. Then, it is natural to assume $\underline{u}(\underline{x}_o) = \lim_{n\to\infty} \underline{u}_n(\underline{x}_o)$ where

$$\underline{u}_n(\underline{x}_o) = \int (\nabla^{\perp}_{x} g_D)(\underline{x}_o,\underline{y})\omega_n(\underline{y})dy =$$

$$= \underline{\nabla}^{\perp}\int -\frac{1}{2\pi}\ln|\underline{x}_o-\underline{y}|\omega_n(\underline{y})d\underline{y} + \underline{\nabla}^{\perp}\int \gamma_D(\underline{x}_o,\underline{y})\omega_n(\underline{y})d\underline{y}\ . \tag{2.6}$$

Since ω_n is spherically symmetric and $\gamma_D(\underline{x},\underline{y})$ is symmetric in

$\underline{x},\underline{y}$, then:

$$\lim_{n\to\infty} \underline{u}_n(\underline{x}_o) = \underline{\nabla}_{\underline{y}}^{\perp}\gamma_D(\underline{y},\underline{x}_o)\Big|_{\underline{y}=\underline{x}_o} = \frac{1}{2}\underline{\nabla}_{\underline{x}_o}^{\perp}\tilde{\gamma}_D(\underline{x}_o) \tag{2.7}$$

where

$$\tilde{\gamma}_D(\underline{x}) = \gamma_D(\underline{x},\underline{x}) \tag{2.8}$$

If $D = \mathbb{R}^2$, $\tilde{\gamma}_{\mathbb{R}^2} = 0$ and this says that a single vortex does not move in its own field. In presence of walls, a single vortex moves under the action of the boundary. This is not surprising: a smoke ring, that by symmetry may be thought as a single vortex in a half plane, moves along its symmetry axis.

Now we state the model by writing the evolution equation describing the dynamics of the vortices:

$$\dot{\underline{x}}_i = \underline{\nabla}_i^{\perp} \sum_{\substack{j=1\\ j\neq i}}^{n} a_j g_D(\underline{x}_i,\underline{x}_j) + \frac{1}{2}\underline{\nabla}_i^{\perp} a_i \tilde{\gamma}_D(\underline{x}_i) . \tag{2.9}$$

Defining

$$\omega_t^n(d\underline{x}) = \sum_{i=1}^{n} a_i \delta_{\underline{x}_i(t)}(d\underline{x}) \tag{2.10}$$

where $\{\underline{x}_i(t)\}_{i=1}^n$ is the solution of (2.9) with initial condition $\{\underline{x}_i\}_{i=1}^n$, we have, by direct computation,

$$\frac{d}{dt}\omega_t^n(f) = \omega_t^n(\underline{u}_o\cdot\underline{\nabla}f) \tag{2.11}$$

where

$$\underline{u}_o(\underline{x},t) = \int_D \nabla^{\perp}_{\underline{x}} g_D(\underline{x},\underline{x}')\chi(\{\underline{x} \neq \underline{x}'\})\omega^n_t(d\underline{x}')$$

$$+ \frac{1}{2}\int_D \nabla^{\perp}\tilde{\gamma}(\underline{x})(1-\chi(\{\underline{x} \neq \underline{x}'\}))\omega^n_t(d\underline{x}') \tag{2.12}$$

and $\chi(\{\ \})$ denotes the indicator of the set $\{\ \}$.

We remark that the model we have introduced is equivalent to the Euler equations, assuming as weak form of the Euler equations (2.11) and (2.12), being equivalent to the usual Euler equations (2.4) in case of sufficiently smooth ω.

Obviously, the euristic justifications we have given for the vortex model (2.9), are not satisfactory.

The symmetry argument (2.6), (2.7) holds at time zero only, and, in some sense, is misleading. We discuss this point in details in Section 4, after that some mathematical aspects of Euler evolution will be investigated. From now on we shall assume the vortex model and discuss some of its properties, without further criticisms on its fundaments.

We start by observing that the model is Hamiltonian. Eq.s (2.9) can be written as

$$\begin{cases} a_i\dot{x}_{i1} = \dfrac{\partial H}{\partial x_{i2}} \\ a_i\dot{x}_{i2} = -\dfrac{\partial H}{\partial x_{i1}} \end{cases} \tag{2.13}$$

where $\dot{x}_{i\alpha}$, $\alpha = 1,2$, denote the component of $\underline{x}_i$ and

$$H = \frac{1}{2}\sum_{i \neq j} a_i a_j g_D(\underline{x}_i,\underline{x}_j) + \frac{1}{2}\sum_i a_i^2 \tilde{\gamma}_D(\underline{x}_i) \tag{2.14}$$

we remark that the conjugate variables are $\sqrt{|a_i|}\,x_{i1}$ and

$\sqrt{|a_i|}\, x_{i2} \operatorname{Sgn} a_i$.

Obviously the energy H is a constant of motion and Liouville theorem holds. Moreover, if H is translationally invariant, then

$$\underline{M} = \sum_{i=1}^{n} a_i \underline{x}_i = \text{const.} \tag{2.15}$$

If H is rotationally invariant, then

$$I = \sum_{i=1}^{n} a_i |\underline{x}_i|^2 = \text{const.} \tag{2.16}$$

Notice that $\underline{M} = (\sum_{i=1}^{n} a_i)\underline{B}$, where $\underline{B}$ is the center of vorticity of the system.

When dealing with two vortices in $\mathbb{R}^2$, the above first integrals are sufficient to determine the motion completely.

If $a_1 \neq -a_2$, they rotate with uniform speed around the center of vorticity. If $a_1 = -a_2$ they go on parallel straight lines with uniform velocity.

The Hamiltonian systems we are considering are non integrable, except in particular situations. We shall discuss shortly this important point later.

A preliminary problem in the study of the qualitative behavior of this dynamical system, is the existence and uniqueness of the solutions. This problem is not trivial, since the logarithmic divergence of the Green function may generate catastrophes, not prevented by the first integrals of motion. A case in which one can exclude singularities in finite time, is the motion of a completely positive (or negative) vortex system in $\mathbb{R}^2$. In this case energy and second moment (2.16)) conservations, exclude a vortex to have infinite velocity in a finite time. However, singularities may arise in general situa tions. Let us give an example.

Consider $D = \mathbb{R}^2$, $n = 3$, $a_1 = 2$, $a_2 = 2$, $a_3 = -1$, $\underline{x}_1 = (-1,0)$, $\underline{x}_2 = (1,0)$, $\underline{x}_3 = (1,\sqrt{2})$, and evolve this system accordingly to Eq.s (2.9) that reduce to

$$\frac{d}{dt} l_{ij}^2 = \frac{2}{\pi} a_k A\left(\frac{1}{l_{jk}^2} - \frac{1}{l_{ki}^2}\right) \tag{2.17}$$

where $l_{ij} = |\underline{x}_i - \underline{x}_j|$, i,j,k are the names of the three vortices appearing in the counter clockwise order, and A is the area of the $\underline{x}_1$, $\underline{x}_2$, $\underline{x}_3$ triangle. It is easily seen that the ratios bet ween the sides are conserved during the motion and

$$\frac{d}{dt} l_{ij}^2(t) = -\frac{1}{3\sqrt{2}\pi} l_{ij}^2(0) \tag{2.18}$$

and hence

$$l_{ij}(t) = l_{ij}(0)\sqrt{1 - \frac{t}{3\sqrt{2}\pi}} \tag{2.19}$$

Therefore $l_{ij} = 0$ if $t = 3\sqrt{2}\pi$. (See fig. 1).

The occurence of collapses excludes the possibility of giving an existence and uniqueness theorem for solutions of the system (2.9), for all times and for all initial conditions. What one can hope is that collapses are "exceptional".

A related problem, perhaps more interesting from a physical point of view, is to establish the occurrence of ε-collapses i.e. events in which two vortices are at distance less than ε. This because it is not experimentally possible to distinguish between a mathematical collapse and an ε-collapse, when ε is small enough.

We assume the Lebesgue measure to be the measure of exceptionality of an event. Then the above problem reduces to estimate the probability (w.r.t. the normalized Lebesgue measure) for are ε-collapse to occur. If such a probability is infinitesimal in ε, then collapses do not take place with probability one and a global flow may be constructed.

We prove this feature in the simple case in which D is a circular domain of radius R. Then

$$g_D(\underline{x},\underline{y}) = -\frac{1}{2\pi}\ln\frac{|\underline{x}-\underline{y}|R}{|\underline{y}||\underline{x}-\bar{\underline{y}}|} \tag{2.20}$$

where

$$\bar{\underline{y}} = \left(\frac{R^2 y_1}{|\underline{y}|^2}, \frac{R^2 y_2}{|\underline{y}|^2}\right), \quad \underline{y} = (y_1, y_2) \tag{2.21}$$

is the conjugate point of $\underline{y}$.

The key of the proof is that the Lebesgue measure is formally invariant for the flow and this leads to time zero estimate.

Consider a regularization of the Green function g_D given by a C^∞ (D × D) function defined as

$$g_\varepsilon(\underline{x},\underline{y}) = -\frac{1}{2\pi}\ln_\varepsilon \frac{|\underline{x}-\underline{y}|R}{|\underline{y}||\underline{x}-\bar{\underline{y}}|} \tag{2.22}$$

where $\ln_\varepsilon : \mathbf{R} \to \mathbf{R}$ is an even C^∞ function not decreasing in $[0,+\infty)$ and such that ($\varepsilon < 1$):

$$\begin{aligned} &\ln_\varepsilon x = \ln|x| && \text{if } |x| > \varepsilon \\ &|\ln_\varepsilon x| \leq |\ln|x|| && x \in \mathbf{R} \\ &\left|\frac{d}{dx}\ln_\varepsilon x\right| \leq \frac{1}{|x|} && x \in \mathbf{R}. \end{aligned} \tag{2.23}$$

We define also

$$\tilde{\gamma}_\varepsilon(\underline{x}) = \frac{1}{2\pi}\ln_\varepsilon\left(\frac{R^2 - |\underline{x}|^2}{R}\right) \tag{2.24}$$

By construction g_ε and $\tilde{\gamma}_\varepsilon$ converge to g_D and $\tilde{\gamma}_D$ as $\varepsilon \to 0$ point wese.

We notice that if $|\underline{x}-\underline{y}| > 3R\varepsilon$ then $g_\varepsilon(\underline{x},\underline{y}) = g_D(x,y)$ and if $|x| < R-\varepsilon$ then $\gamma_\varepsilon(\underline{x}) = \gamma_D(\underline{x})$.

We define the regularized vortex dynamics

$$\dot{\underline{x}}_i^\varepsilon(t) = \sum_{\substack{j=1\\ j\neq i}}^{N} a_j \underline{\nabla}_i^\perp g_\varepsilon(\underline{x}_i^\varepsilon(t),\underline{x}_j^\varepsilon(t)) + \frac{a_i}{2}\underline{\nabla}_i^\perp \tilde{\gamma}_\varepsilon(\underline{x}_i(t)) \tag{2.25}$$

$$x_i^\varepsilon(0) = x_i, \qquad X = \{\underline{x}_i\}_{i=1}^N \in D^N$$

The above dynamics is well defined as a flow $S_t^\varepsilon: D^N \to D^N$: $S_t^\varepsilon X \equiv X^\varepsilon(t) = \{\underline{x}_1^\varepsilon(t).....\underline{x}_N^\varepsilon(t)\} \in D^N$ since $\underline{\nabla}^\perp g_\varepsilon$ and $\underline{\nabla}^\perp\tilde{\gamma}_\varepsilon$ are smooth functions satisfying $\underline{\nabla}^\perp g_\varepsilon \cdot n = \underline{\nabla}^\perp \tilde{\gamma}_\varepsilon \cdot n = 0$. The last property prevents vortices to hit the wall. Moreover the flow (2.25) is hamiltonian and hence preserves the Lebesgue measure.

We define

$$d_T^\varepsilon(X) = \min\left(\min_{i\neq j}\ \inf_{t\in[0,T]} |\underline{x}_i^\varepsilon(t)-\underline{x}_j^\varepsilon(t)|, \min_i\ \inf_{t\in[0,T]} R-|\underline{x}_i(t)|\right) \tag{2.26}$$

and $\bar{\varepsilon} = \max(\varepsilon, 3R\varepsilon)$.

<u>Theorem 2.1</u>. Let $\lambda(dX) \equiv \dfrac{d\underline{x}_1.....d\underline{x}_N}{(\pi R^2)^N}$ be the normalized Lebesgue measure. Then for $\varepsilon < 1$, and arbitrary $\delta \in (0,1)$, there exists a constant A depending only on δ,R,N such that:

$$\lambda(\{X \in D^N |\ d_T^\varepsilon(X) < \bar{\varepsilon}\}) \leq A(T+1)\varepsilon^{1-\delta}\ . \tag{2.27}$$

Proof. Let us define

$$\varphi_\varepsilon(X) = \frac{1}{2} \sum_{i\neq j} F(g_\varepsilon(\underline{x}_i,\underline{x}_j)) + \sum_i F(-\tilde{\gamma}_\varepsilon(\underline{x}_i)) \tag{2.28}$$

where $F(r) = \exp(1-\eta)r$ and $1 > \eta > 1-2\pi$. We have:

$$\dot{\varphi}_\varepsilon(S_t^\varepsilon X) = \sum_{i\neq j} F'(g_\varepsilon(\underline{x}_i^\varepsilon(t),\underline{x}_j^\varepsilon(t)))(\underline{\nabla}_i g_\varepsilon)(\underline{x}_i^\varepsilon(t),\underline{x}_j^\varepsilon(t))\cdot\dot{\underline{x}}_i^\varepsilon(t)$$
$$- \sum_i F'(-\tilde{\gamma}_\varepsilon(\underline{x}_i))(\underline{\nabla}_i\tilde{\gamma}_\varepsilon)(\underline{x}_i^\varepsilon(t))\cdot\dot{\underline{x}}_i^\varepsilon(t) \tag{2.29}$$

In virtue of (2.25) and the identity:

$$\underline{\nabla} f(\underline{x})\cdot\underline{\nabla}^\perp f(\underline{x}) = 0, \qquad \forall\, f\in C_1(\mathbb{R}^2), \tag{2.30}$$

we have

$$|\dot{\varphi}_\varepsilon(S_t^\varepsilon X)| \leq a \sum_{i=1}^N \sum_{\substack{j=1\\ j\neq i}}^N \sum_{\substack{k=1\\ k\neq j\\ k\neq i}}^N |F'(g_\varepsilon(\underline{x}_i^\varepsilon(t),\underline{x}_j^\varepsilon(t)))||\underline{\nabla}_i g_\varepsilon(\underline{x}_i^\varepsilon(t),\underline{x}_j^\varepsilon(t))|$$
$$\cdot\ |\underline{\nabla}_i g_\varepsilon(\underline{x}_i^\varepsilon(t),\underline{x}_k^\varepsilon(t))|$$
$$+ a \sum_{\substack{i=1\\ j\neq i}}^N |F'(g_\varepsilon(\underline{x}_i^\varepsilon(t),\underline{x}_j^\varepsilon(t)))||\underline{\nabla}_i g_\varepsilon(\underline{x}_i^\varepsilon(t),\underline{x}_j^\varepsilon(t))\cdot\underline{\nabla}_i^\perp\tilde{\gamma}_\varepsilon(x_i(t))|$$
$$+ a \sum_{\substack{i=1\\ j\neq i}}^N |F'(-\tilde{\gamma}_\varepsilon(\underline{x}_i^\varepsilon(t)))||\underline{\nabla}_i\tilde{\gamma}_\varepsilon(\underline{x}_i^\varepsilon(t))\cdot\underline{\nabla}_i^\perp g_\varepsilon(\underline{x}_i^\varepsilon(t),\underline{x}_j^\varepsilon(t))| \tag{2.31}$$

when $a = \max_{1\leq i\leq N} |a_i|$.

We denote $h = h_1 + h_2 + h_3$, where:

$$\begin{cases} h_1(X) = a \sum_{i=1}^{N} \sum_{\substack{j=1 \\ j\neq i}}^{N} \sum_{\substack{k=1 \\ k\neq j \\ k\neq i}}^{N} |F'(g_\varepsilon(\underline{x}_i,\underline{x}_j))| \, |\underline{\nabla}_i g_\varepsilon(\underline{x}_i,\underline{x}_j)| \, |\underline{\nabla}_i g_\varepsilon(\underline{x}_i,\underline{x}_k)| \\ h_2(X) = a \sum_{i\neq j} |F'(g_\varepsilon(\underline{x}_i,\underline{x}_j))| \, |(\underline{\nabla}_i g_\varepsilon)(\underline{x}_i,\underline{x}_j)\cdot\underline{\nabla}_i^{\perp}\cdot\tilde{\gamma}_\varepsilon(\underline{x}_i)| \\ h_3(X) = a \sum_{i\neq j} |F'(-\tilde{\gamma}_\varepsilon(\underline{x}_i))| \, |\underline{\nabla}_i\tilde{\gamma}_\varepsilon(\underline{x}_i)\cdot\nabla_i^{\perp} g_\varepsilon(\underline{x}_i,\underline{x}_j)| \end{cases} \tag{2.32}$$

We now want to show that there exists a constant C not depending on ε such that

$$\int h_i(X)\lambda(dX) < C < +\infty, \qquad i = 1,2,3. \tag{2.33}$$

Since

$$|g_\varepsilon(\underline{x},\underline{y})| \leq \frac{1}{2\pi}\ln\frac{2R}{|\underline{x}-\underline{y}|} \tag{2.34}$$

$$|\underline{\nabla} g_\varepsilon(\underline{x},\underline{y})| \leq \frac{\text{Const.}}{|\underline{x}-\underline{y}|}$$

$$h_1(X) \leq \text{Const.} \sum_i \sum_{j\neq i} \sum_{\substack{k\neq j \\ k\neq i}} \frac{1}{|\underline{x}_i-\underline{x}_j|^{(\frac{1-\eta}{2\pi})+1}} \cdot \frac{1}{|\underline{x}_i-\underline{x}_k|} \tag{2.35}$$

hence (2.33) is proved with $i = 1$.

To estimate h_2 and h_3 it will be convenient to use the following inequality:

$$|(\nabla_{\underline{x}} g_\varepsilon)(\underline{x},\underline{y})\cdot\underline{\nabla}\tilde{\gamma}_\varepsilon(\underline{x})| \leq \frac{M}{|\underline{x}-\underline{y}|(|\underline{x}-\underline{y}| + \text{dist}(\underline{x},\partial D))} \tag{2.36}$$

where M depends only on R. Estimate (2.36) will be proved in the Appendix at the end of the Section. By the use of (2.36) the estimates of h_1 and h_2 reduce to the evaluations of the integrals

$$\int_{D\times D} d\underline{x}d\underline{y}\frac{1}{|\underline{x}-\underline{y}|^{\alpha}(|\underline{x}-\underline{y}|+\mathrm{dist}(\underline{x},\partial D))} \tag{2.37}$$

$$\int_{D\times D} d\underline{x}d\underline{y}\frac{1}{|\underline{x}-\underline{y}|(|\underline{x}-\underline{y}|+\mathrm{dist}(\underline{x},\partial D))\mathrm{dist}(\underline{x},\partial D)^{\beta}} \tag{2.38}$$

where $\alpha < 2$ and $\beta < 1$. Putting $(\underline{x}-\underline{y})_1 = \rho\cos\theta$, $(\underline{x}-\underline{y})_2 = \rho\sin\theta$, $x_1 = \xi\cos\varphi$, $x_2 = \xi\sin\varphi$, we obtain

$$(2.37) \leq \mathrm{Const.}\int_0^{2R} d\rho\int_0^R d\xi\,\frac{\rho\xi}{\rho^{\alpha}(\rho+(R-\xi))} < +\infty$$

$$(2.38) \leq \mathrm{Const.}\int_0^{2R} d\rho\int_0^R d\xi\,\frac{\xi}{(\rho+(R-\xi))(R-\xi)^{\beta}} < +\infty \tag{2.39}$$

This complete the proof of (2.33).

We notice also that $\int|\varphi_\varepsilon|d\lambda$ is bounded uniformly in ε. Therefore:

$$\int_{D^N}\lambda(dX)\sup_{0\leq t\leq T}|\varphi_\varepsilon(S_t^\varepsilon X)| \leq \int_{D^N}\lambda(dX)|\varphi_\varepsilon(X)|+\int_{D^N}\lambda(dX)\int_0^T dt|\dot{\varphi}(S_t^\varepsilon X)|$$

$$\leq \int_{D^N}\lambda(dX)|\varphi_\varepsilon(X)|+\int_{D^N}\lambda(dX)\int_0^T dt|h(S_t^\varepsilon X)| = \int_{D^N}\lambda(dX)|\varphi_\varepsilon(X)|+T\int_{D^N}\lambda(dX)h(X) \tag{2.40}$$

(By Fubini theorem and the time invariance of the measure λ)

$\leq A_0(1+T)$,

where A_o is a positive constant depending only on R, N and η. Since $\{X|\ d_T^\varepsilon(X) < \bar{\varepsilon}\} \subset \{X|\ \varphi_\varepsilon(X) \geq F(-\frac{1}{2\pi} \ln 2\bar{\varepsilon})\}$, by Chebychev's inequality it follows

$$\lambda(\{X|\ d_T^\varepsilon(X) < \bar{\varepsilon}\}) \leq A_o(1+T)\bar{\varepsilon}^{(\frac{1-\eta}{2\pi})} \tag{2.41}$$

and hence the thesis. □

In virtue of this Theorem a global flow λ-a.e. defined can be constructed by putting $S_tX = S_t^\varepsilon X$ $0 \leq t \leq T$ if $d_T^\varepsilon(X) > \bar{\varepsilon}$, for all $\varepsilon > 0$. Obviously S_tX is the (unique) solution of the vortex dynamics (2.9).

A similar result has been obtained in case in which $D = T^2$. The proof is simpler because of the absence of the boundaries [Dül].

In case of general bounded domains,the interaction vortex-boundary gives rise to a term that have the same singularity as in case of circular domains in which everything is explicit. Thus we believe that the above proof, combined with suitable estimates on the Green function, works also in general.

We discuss now the case $D = \mathbf{R}^2$. Here the previous proof cannot work because of the non compactness of D. In fact a vortex could, in principle, arrive at infinity in finite time and to avoid this feature new ideas are necessary.

We shall show, under an additional hypothesis and using the conservation of the center of vorticity, that a system of vortices remains confined in bounded regions in finite time. Then combining this argument with the previous one, we establish existence and uniqueness for the initial value problem of a vortex system in $\mathbf{R}^2$.

Assume

$$\sum_{i \in P(N)} a_i \neq 0 \tag{2.42}$$

where $P(N)$ indicates any subset of $\{1.....N\}$, and consider the set of all initial conditions for which $|\underline{x}_i| \leq n$.

We want to prove the existence of a constant C depending on N, n, T, $a_1.....a_N$ and independent of ε (and the initial conditions) for which

$$\sup_{1 \leq i \leq N} \sup_{0 \leq t \leq T} |\underline{x}_i^{\varepsilon}(t) - \underline{x}_i| \leq C \tag{2.43}$$

where $\underline{x}_i \to \underline{x}_i^{\varepsilon}(t)$ is the solution of the regularized dynamics introduced before.

We shall prove (2.43) by induction.

Assume $\underline{\tilde{x}}_i^{\varepsilon,k}(t)$ be the solution of the initial value problem involving k vortices starting at time zero from the points $\underline{x}_i$, under the additional action of an external field of norm less than 1. If there exists $C(k) > 0$ (independent of ε) such that

$$\sup_{i} \sup_{0 \leq t \leq T} |\underline{\tilde{x}}_i^{\varepsilon,h}(t) - \underline{x}_i| \leq C(k) , \quad h \leq k \tag{2.44}$$

then we can find $C(k+1) > C(k)$ (also independent of ε) such that:

$$\sup_{i} \sup_{0 \leq t \leq T} |\underline{\tilde{x}}_i^{\varepsilon,k+1}(t) - \underline{x}_i| \equiv R \leq C(k+1) . \tag{2.45}$$

Since (2.44) is trivially verified if $k = 1$ (with $C(1) = T$) (2.43) will follow by (2.45) by putting $C = C(N)$.

Denoting:

$$\begin{cases} a = \max\limits_{1\le i\le N} |a_i| , \quad A = \min\limits_{P(N)} \left| \sum\limits_{i\in P(N)} a_i \right| \\ \underline{M}_k(t) = \sum\limits_{j=1}^{k} a_j \underline{x}_j(t) , \end{cases} \tag{2.46}$$

we have:

$$\begin{aligned} |\underline{M}_{k+1}(t)| &= \left|\sum_{j=1}^{k+1} a_j \underline{x}_j(t)\right| \le |\underline{M}_{k+1}(0)| + T(k+1) \\ &\le (k+1)a(n+\frac{T}{a}) \le (N+1)a(n+\frac{T}{a}) \equiv b \end{aligned} \tag{2.47}$$

Suppose

$$R > A^{-1}(b + nA + N^2 a[2C(k) + aN + 2n]) \tag{2.48}$$

Then for some index i_1 and some time $t^* \le T$:

$$|\underline{x}_{i_1} - \underline{\tilde{x}}_{i_1}^{\varepsilon,k+1}(t^*)| = R. \tag{2.49}$$

Hence

$$| \underline{M}_{k+1}(t^*)| = \left|\sum_{j=1}^{k+1} a_j \underline{\tilde{x}}_{i_1}^{\varepsilon,k+1}(t^*) + \sum_{j=1}^{k+1} a_j \underline{y}_j(t^*)\right| \le b \tag{2.50}$$

where

$$\underline{y}_j(t^*) = \underline{\tilde{x}}_j^{\varepsilon,k+1}(t^*) - \underline{\tilde{x}}_{i_1}^{\varepsilon,k+1}(t^*). \tag{2.51}$$

From (2.50) there exists i_2 such that:

$$|\underline{y}_{i_2}| \ge \frac{(R-n)A - b}{aN} \tag{2.52}$$

Therefore at time t_1^* the vortices can be divided in two clusters at distance larger than $d = \frac{(R-n)A-b}{aN^2}$, and the field produced by each cluster on the other is smaller then $\frac{1}{2\pi}\frac{Na}{d} = \frac{1}{2\pi}\frac{a^2N^3}{(R-n)A-b} < 1$. One realizes easily, using the inductive hypotheses, that these two clusters are at distance larger then $d' = d - 2C(k)$ for all time $t \leq T$. Hence $d' > n$ in particular for $t = 0$, which yields a contraddiction. Thus inequality (2.48) has to be reversed and (2.45) is proved with $C(k+1)$ given by the r.h.s. of (2.48).

The dynamics of a N vortex system satisfying $|\underline{x}_i| \leq n$ $i = 1.....N$ can now be constructed using the same idea of the compact case and observing that the region $\{\underline{x}_1^\varepsilon(t)...\underline{x}_N^\varepsilon(t) \mid |\underline{x}_i| \leq n,\ 0 \leq t \leq T\}$ is enclosed in a sphere in $\mathbb{R}^{2N}$ of a radius independent of ε. Hence the set of initial configurations exhibiting singular behaviour have Lebesgue measure zero. The union on all positive integers n of such sets have still Lebesgue measure zero and this achieves the proof.

Let us come back to the qualitative behavior of the vortex motion. Consider the case of three vortices in T^2. The three first integrals $\underline{M} \equiv (M_1, M_2)$ and H (see def. 2.15) are independent and in involution. In fact a simple calculation shows that the gradients are independent and that $\{M_1, M_2\} = \{M_i, H\} = 0$ where $\{\cdot,\cdot\}$ denotes, as usual, the Poisson Brackets (recall that, by circulation theorem, $\sum a_i = 0$).

Liouville theorem, [Arn.1] suggests that this system is integrable.

It is also known that a system of three vortices is completely integrable in $\mathbb{R}^2$. A detailed description of this motion is given in [No.1] and [Ar.1]. Nevertheless the situation may considerably be different for a test vortex of null vorticity, or what is the same, for the motion of a molecule of the fluid. Such motion, for suitable values of a_i's, might be chaotic: the text vortex could be very often scattered by other vortices in a high energy situation. Numerical experiments seem to confirm this behavior [Ar.2].

It has been proved by Ziglin [Kh.1], that the motion of four vortices is not integrable. On the other hand in the same paper, Khanin was able to exhibit a positive measure set of initial conditions, for which a quasiperiodic motion takes place. The main idea is the following. Consider the four vortices $\{a_i, \underline{x}_i\}_{i=1}^{4}$ in such a way that x_1 and x_2 are very far apart from x_3 and x_4.

Then if $a_1 \neq -a_2$, $a_3 \neq -a_4$, $a_1 + a_2 \neq a_3 + a_4$, the global hamiltonian H can be written as $H_o + V$, where H_o is the interaction of the two pairs of vortices and the interaction of their center of vorticity, and $V = H - H_o$ (small perturbation if the two pairs are for enough). The thesis follows by an application of the KAM theorem [KAM1], since the unperturbed motion given by H_o is quasiperiodic. This result can be generalized to an arbitrary number of vortices by induction. See fig. 2 for the motion of two weakly interacting pairs of two vortices.

We can extend this result, by the use of the same idea, for a system moving in a bounded region.

In this case we cannot arrange the vortices in very far clusters because of the boundness of the region. Nevertheless we can use scaling properties of the interaction to construct weakly interacting clusters.

To give the idea we study the case of three identical vortices of intensity one in a circular domain. More general cases can be treated along the same lines.

Let us write the equations of motion

$$\begin{cases} \dot{x}_i = \dfrac{\partial H}{\partial y_i} \\ \dot{y}_i = -\dfrac{\partial H}{\partial x_i} \end{cases} \qquad \sqrt{x_i^2 + y_i^2} < 1, \quad i = 1,2,3, \quad \underline{x}_i = (x_i, y_i). \tag{2.53}$$

The Hamiltonian is given by:

$$H = -\frac{1}{2\pi} \sum_{\substack{i,j=1 \\ i<j}}^{3} \ln \frac{|\underline{x}_i - \underline{x}_j|}{|\underline{x}_j| \, \left|\underline{x}_i - \dfrac{\underline{x}_i}{|\underline{x}_j|^2}\right|} + \frac{1}{4\pi} \sum_{i=1}^{3} \ln (1 - |\underline{x}_i|^2) . \tag{2.54}$$

We introduce the change of scale

$$\underline{x}_i = \frac{1}{\alpha} \underline{z}_i \qquad t = \frac{\tau}{\alpha^2} \qquad \alpha > 0 \tag{2.55}$$

after which, the motion eq.s reads as

$$\frac{d}{d\tau} \underline{z}_i = \nabla_z^{\perp} \tilde{H} \tag{2.56}$$

where

$$\tilde{H} = -\frac{1}{2\pi} \sum_{\substack{i,j=1 \\ i<j}}^{3} \ln \frac{|\underline{z}_i - \underline{z}_j|}{|\underline{z}_j| \, \left|\dfrac{\underline{z}_i}{\alpha} - \dfrac{\underline{z}_j \alpha}{|\underline{z}_j|^2}\right|} + \frac{1}{4\pi} \sum_{i=1}^{3} \ln \left(1 - \frac{|z_i|^2}{\alpha^2}\right),$$

$$|\underline{z}_i| \leq \alpha \tag{2.57}$$

Introducing the canonical transformation ($\underline{z}_i = (z_{i,1}; z_{i,2})$)

$$
\begin{aligned}
z_{1,1} &= \sqrt{\tfrac{2}{3}}\,\sqrt{p_3}\,\cos q_3 + \tfrac{1}{\sqrt{3}}\sqrt{p_2}\,\cos q_2 + \sqrt{p_1}\,\cos q_1 \\
z_{1,2} &= \sqrt{\tfrac{2}{3}}\,\sqrt{p_3}\,\sin q_3 + \tfrac{1}{\sqrt{3}}\sqrt{p_2}\,\sin q_2 + \sqrt{p_1}\,\sin q_1 \\
z_{2,1} &= \sqrt{\tfrac{2}{3}}\,\sqrt{p_3}\,\cos q_3 + \tfrac{1}{\sqrt{3}}\sqrt{p_2}\,\cos q_2 + \sqrt{p_1}\,\cos q_1 \\
z_{2,2} &= \sqrt{\tfrac{2}{3}}\,\sqrt{p_3}\,\sin q_3 + \tfrac{1}{\sqrt{3}}\sqrt{p_2}\,\cos q_2 - \sqrt{p_1}\,\sin q_1 \\
z_{3,1} &= \sqrt{\tfrac{2}{3}}\,\sqrt{p_3}\,\cos q_3 - \tfrac{2}{\sqrt{3}}\sqrt{p_2}\,\cos q_2 \\
z_{3,2} &= \sqrt{\tfrac{2}{3}}\,\sqrt{p_3}\,\sin q_3 - \tfrac{2}{\sqrt{3}}\sqrt{p_2}\,\sin q_2 \ .
\end{aligned}
\tag{2.58}
$$

$\sqrt{p_1}$, $\sqrt{p_2}$, $\sqrt{p_3}$ are proportional to $|\underline{z}_1 - \underline{z}_2|$, $|\underline{z}_3 - \frac{\underline{z}_1+\underline{z}_2}{2}|$, $|\frac{\underline{z}_1+\underline{z}_2+\underline{z}_3}{3}|$, and q_1 q_2 q_3 are the angular variables associated to the action variables, that are proportional to the logarithms of the p variables.

The unperturbed Hamiltonian takes the form:

$$\tilde{H}_o = -\frac{1}{4\pi}\ln p_1 - \frac{1}{4\pi}\ln p_2 + \frac{9}{4\pi}\ln\left(1 - \frac{3}{2}\tilde{p}_3\right) \tag{2.59}$$

where $\tilde{p}_3 = p_3/\alpha^2$.

By direct computation

$$\tilde{H} - \tilde{H}_o = \frac{3}{\pi}\ln\alpha + \text{terms vanishing when } p_2 \text{ and } \alpha \text{ are large} \tag{2.60}$$

Quasi-periodic motions follow again by KAM theorem. (See fig. 3 illustrating the motion of a vortex pair weakly interacting with a third vortex in a circular domain)

Notes

Vortex theory was introduced by Helmholtz one century ago [He.1]. Kirchoff [Ki.1] and Poincaré [Po.1] outlined the Hamiltonian structure of the system. The example of vortex collapse is taken by [Ar.1].

For a modern, review of the qualitative aspects of few vortex motion see [Ar.3].

The problem of estimating the breakdown of the solution for divergenceless vector field, with a compact set of singularities, has been studied in [A.1]. For a review concerning the problem for Newtonian system see [Saa.1]. See also [Ma.3].

The Hamiltonian character of the vortex system allows the use of the methods of the Classical Statistical Mechanics. A first study of a vortex gas has been proposed by Onsager [On.1]. He discussed Gibbs states of negative temperature to explain the presence of one sign vortex clusters.

For a rigorous discussion of the Statistical Mechanics of a gas of vortices in the thermodynamical limit, see [Fr.1].

Time evolution of infinitely many vortices in a strip, in a thermodynamical limit has been constructed in [Ma.4].

Appendix

Since

$$|\nabla_{\underline{x}} g_\varepsilon(\underline{x},\underline{y})\cdot\nabla^{\perp}\tilde{\gamma}_\varepsilon(\underline{x})| \leq |\nabla_{\underline{x}} g(\underline{x},\underline{y})\cdot\nabla^{\perp}\tilde{\gamma}(\underline{x})| \qquad (A.1)$$

it is enough to prove (2.36) replacing g_ε and $\tilde{\gamma}_\varepsilon$ by g and γ.

Assuming $\underline{y} = (y,0)$, $y \geq 0$ and $R=1$ for sake of simplicity, we put $\underline{x} = (\rho\cos\theta, \rho\sin\theta)$ and hence

$$|\nabla_{\underline{x}} g(\underline{x},\underline{y})\cdot\underline{\nabla}^{\perp}\tilde{\gamma}(\underline{x})| = \frac{1}{4\pi^2}\frac{1}{\rho}\left|\frac{\partial}{\partial\theta}\ln\frac{\rho^2+y^2-2\rho y\cos\theta}{\rho^2 y^2+1-2\rho y\cos\theta}\,\frac{\partial}{\partial\rho}\ln(1-\rho^2)\right| \qquad (A.2)$$

By straightforward calculation we realize that the bound (2.36) is equivalent in proving the boundness of the expression

$$\rho\left|\frac{\sin\theta y(1+y)(1-y)[\sqrt{(\rho^2+y^2-2\rho y\cos\theta)}+(1-\rho)]}{\sqrt{(\rho^2+y^2-2\rho y\cos\theta)}\;(\rho^2 y^2+1-2\rho y\cos\theta)}\right| \qquad (A.3)$$

$$0 \leq \rho \leq 1, \qquad 0 \leq y \leq 1, \qquad 0 \leq \theta < 2\pi$$

If $y < \frac{1}{2}$

$$(A.3) \leq \text{Const.}\frac{\rho y|\sin\theta|}{\sqrt{(\rho-y)^2+2(1-\cos\theta)\rho y}} \leq \text{Const.}\frac{\rho y|\sin\theta|}{\sqrt{\rho y}\,\sqrt{1-\cos\theta}} < +\infty \qquad (A.4)$$

If $y \geq \frac{1}{2}$ and $\rho \leq \frac{1}{3}$ (A.3) is manifestly bounded.

Finally if $y \geq \frac{1}{2}$ and $\rho \geq \frac{1}{3}$ we have

$$(A.3) \leq \text{Const.}\{B_1 + B_2\} \qquad (A.5)$$

where

$$B_1 = \frac{|\sin\theta|(1-y)}{(1-\rho y)^2+2(1-\cos\theta)\rho y} \tag{A.6}$$

$$B_2 = \frac{|\sin\theta|(1-y)(1-\rho)}{\sqrt{(\rho-y)^2+2\rho y(1-\cos\theta)}\;[(1-\rho y)^2+2(1-\cos\theta)\rho y]} \tag{A.7}$$

Then

$$B_1 \leq \frac{|\sin\theta|(1-y)}{2(1-\rho y)\sqrt{2(1-\cos\theta)\rho y}} \leq \text{Const}\,\frac{(1-y)}{(1-\rho y)} < +\infty \tag{A.8}$$

Finally

$$B_2 \leq \text{Const}\,\frac{|\sin\theta|}{\sqrt{1-\cos\theta}}\cdot\frac{(1-y)(1-\rho)}{(1-\rho y)^2} < +\infty \tag{A.9}$$

3. AN EXISTENCE THEOREM FOR EULER EQUATIONS

In this section we shall establish an existence and uniqueness theorem for the Euler evolution, convenient for what follows and interesting in itself. The basic idea is well known and is based on the method of characteristics.

An inspection of Eq. $(1.19)_a$ suggests that if the time dependent vector field $\underline{u}(\underline{x},t)$ would be known, the solution $\omega(\underline{x},t)$ could be expressed in terms of the solution of an ordinary differential equation. More precisely, let $V_t\underline{x}$ be the solution of the initial value problem

$$\begin{cases} \dfrac{dV_t}{dt}\underline{x} = \underline{u}(V_t\underline{x}, t) \\ V_o\underline{x} = \underline{x} \end{cases} \tag{3.1}$$

then

$$\omega(\underline{x},t) = \omega(V_{-t}\underline{x},0) \; . \tag{3.2}$$

Thus the problem of the existence reduces in matching $\underline{u}$ with the velocity field computed from (3.2) via $(1.19)_b$.

Since we are interested in the connection between vortex theory and Euler evolution, it is convenient to study the Euler equation in the weak form stated in Section 2.

The main technical problem in performing our program is due to the singularity of the integral kernel $\underline{\nabla}^{\perp}g_D$. It is natural to introduce a regular version of it. The way we regularize $\underline{\nabla}^{\perp}g_D$, has a physical meaning and preserves the boundary conditions.

Let g_D be the fundamental solution of the Poisson equation with Dirichlet boundary conditions. We define

$$g_\varepsilon(\underline{x},\underline{y}) = \int_D \rho_\varepsilon(|\underline{z}-\underline{y}|)g_D(\underline{x},\underline{z})d\underline{z} \tag{3.3}$$

when $\rho_\varepsilon \in C^k(\mathbb{R}^1)$, $k \geq 2$.
We define

$$\underline{k}_\varepsilon(\underline{x},\underline{y}) = \underline{\nabla}_{\underline{x}} g_\varepsilon(\underline{x},\underline{y}) \,. \tag{3.4$_1$}$$

In the sequel we need ρ_ε to satisfy the following assumptions:

$$\rho_\varepsilon \geq 0 \,, \quad \int_{\mathbb{R}^2} \rho_\varepsilon(|\underline{x}|)d\underline{x} = 1 \,, \quad \text{supp}\, \rho_\varepsilon \subset [-\varepsilon,\varepsilon] \,, \tag{3.4$_2$}$$

hence:

$$\int_D \rho_\varepsilon(|\underline{x}-\underline{y}|)f(\underline{y}) \xrightarrow[\varepsilon\to 0]{} f(\underline{x}) \,, \tag{3.4$_3$}$$

f continuous.

Moreover we require (for some C not depending on ε):

$$\begin{aligned} &|g_\varepsilon(\underline{x},\underline{y})| \leq C(1 - \ln|\underline{x}-\underline{y}|) \\ &|\underline{k}_\varepsilon(\underline{x},\underline{y})| \leq \frac{C}{|\underline{x}-\underline{y}|} \\ &\left|\frac{\partial}{\partial x_\alpha} \underline{k}_\varepsilon(\underline{x},\underline{y})\right| \leq \frac{C}{|\underline{x}-\underline{y}|^2} \end{aligned} \tag{3.4$_4$}$$

It is possible to verify that $\rho_\varepsilon(|\underline{x}|) = \frac{k+1}{\pi\varepsilon^2}(1 - (\frac{\underline{x}}{\varepsilon})^2)^k$ if $|x| < \varepsilon$ and $\rho_\varepsilon(|\underline{x}|) = 0$ if $|\underline{x}| \geq \varepsilon$ satisfies all the above assumptions.

We remark that this regularization is different from that used in the previous Section: in this way $g_\varepsilon(\underline{x},\underline{y})$ is no more symmetric but this is not relevant in what follows.

We notice that g_ε is a sort of interaction between two bubbles of vorticity and satisfies the boundary conditions.

With these definitions, a simplified version of the Euler initial value problem can be stated in the following weak form:

$$\begin{cases} \frac{d}{dt}\omega_t^\varepsilon(f) = \omega_t^\varepsilon(\underline{\nabla} f \cdot \underline{u}^\varepsilon) \\ \omega_o^\varepsilon = \omega \end{cases} \tag{3.5}$$

where $f \in C^\infty(D)$, $\omega_t^\varepsilon(f) = \int \omega_t^\varepsilon(d\underline{x})f(\underline{x})$, ω_t^ε one parameter family of signed measures, and finally

$$\underline{u}^\varepsilon(\underline{x},t) = \int \omega_t^\varepsilon(d\underline{y})\underline{k}_\varepsilon(\underline{x},\underline{y}) \ . \tag{3.6}$$

As consequence of (3.6) we have $\underline{u}^\varepsilon \cdot n = 0$ on ∂D where $\underline{n}$ denotes, as usual, the outward normal. Since we look for solution for which $\omega_t^\varepsilon \in L_1(D)$ then (3.6) implies also that $\underline{u}^\varepsilon(\underline{x}) \to 0$ as $|\underline{x}| \to \infty$.

A useful mathematical tool is a distance between signed measures, we are going to define.

Let M be a metric space with bounded metric function $d: M \times M \to \mathbb{R}^+$. We define

$$M(a) = \{\mu \mid \mu \text{ is a Borel measure on } M \text{ s.t. } \int d\mu = a\} \tag{3.7}$$

If $\mu_i \in M(a)$ $i = 1,2$, we denote by $C(\mu_1,\mu_2)$ the set of all joint representations of μ_1 and μ_2, i.e. $\hat{P} \in C(\mu_1,\mu_2)$ if it is a Borel measure on $M \times M$ of total charge a, such that:

$$\int_{M\times M} \hat{P}(dx_1,dx_2)f(x_i) = \int_M \mu_i(dx)f(x) \quad i=1,2 \qquad (3.8)$$

for all bounded measurable functions f.

Then

$$R(\mu_1,\mu_2) = \inf_{\hat{P}\in C(\mu_1,\mu_2)} \int_{M\times M} \hat{P}(dx_1,dx_2)d(x_1,x_2) \qquad (3.9)$$

defines a distance of $M(a)$.

It has been proved (see [Do.2]) that the topology induced by the metric R is equivalent to the weak convergence topology. We recall that a sequence $\{\mu_n\}_{n=1}^{\infty}$ is weakly convergent to μ iff $\mu_n(f) \to \mu(f)$ for all bounded and continuous f.

The metric (3.9) is called Kantorovich-Rubinstein (KR) distance (sometimes Vasershtein distance).

A concrete example giving an idea of the meaning of this distance, is the following.

If $\mu_k(dx) = \frac{a}{N} \sum_{i=1}^{N} \delta_{x_i^k}(dx)$, $k = 1,2$, then

$$R(\mu_1,\mu_2) = \min_{\pi} \frac{a}{N} \sum_{i=1}^{N} d(x_i^1, x_{\pi(i)}^2) \qquad (3.10)$$

where the above minimum is taken over all the permutation π of 1.....N.

The proof of the formula (3.10) will be given in the Appendix A at the end of this Section.

We notice that, by the use of the formula (3.10), one can prove the equivalence of the topology induced by the KR distance with the weak convergence topology, in many cases of interest.

The above notion can be extended to general signed measures. Let $M(a,b)$ be the set of all signed Borel measures on M with a and b the total masses of the positive and negative part according to the Jordan decomposition. If $\mu_i \in M(a,b)$ $i = 1,2$ we define

$$R(\mu_1,\mu_2) = R(\mu_1^+,\mu_2^+) + R(\mu_1^-,\mu_2^-) \tag{3.11}$$

where $\mu_i = \mu_i^+ - \mu_i^-$ is the Jordan decomposition. (3.11) defines a metric on $M(a,b)$ equivalent to the weak convergence topology on $M(a,b)$.

Now we can formulate the results of this section.

Theorem 3.1. Let $\omega \in M(a,b)$, and $D \subseteq \mathbb{R}^2$ a (possibly unbounded) open set with smooth boundary ∂D*. Then there exists a unique function $t \to \omega_t^\varepsilon \in M(a,b)$ satisfying the initial value problem (3.5), (3.6). Moreover, denoting by $V_t^\varepsilon \underline{x}$ the solution of the problem

$$\begin{cases} \frac{d}{dt} V_t^\varepsilon \underline{x} = \underline{u}^\varepsilon(V_t^\varepsilon \underline{x}, t) \\ V_o^\varepsilon \underline{x} = \underline{x}, \quad \underline{x} \in D \end{cases} \tag{3.12}$$

it results

$$\omega_t^\varepsilon(f) = \int \omega^\varepsilon(d\underline{x}) f(V_t \, \underline{x}) \tag{3.13}$$

for all continuous, bounded f.

In proving this theorem (see below) we shall use the KR

*) Sufficiently regular to give sense to the fundamental solution of the Poisson Equation in D.

distance for $M = \mathbb{R}^2$ with the following metric function:

$$d(\underline{x},\underline{y}) = \begin{cases} |\underline{x}-\underline{y}| & \text{if } |\underline{x}-\underline{y}| \leq 1 \\ 1 & \text{otherwise} \end{cases} \tag{3.14}$$

Theorem 3.2. Let the hypotheses of Theorem 3.1 hold. Suppose in addition $\omega(d\underline{x}) = \omega(\underline{x})d\underline{x}$, $\omega \in L_1 \cap L_\infty(D)$. Then there exists a unique function $t \to \omega_t \in L_1 \cap L_\infty(D)$ such that $\omega_t d\underline{x} \in M(a,b)$ satisfying

$$\begin{cases} \frac{d}{dt}\omega_t(f) = \omega_t(\underline{u}\cdot\underline{\nabla}f), & f \in C^\infty(D) \\ \omega_o = \omega \end{cases} \tag{3.15}$$

$$\underline{u}(\underline{x},t) = \int_D (\underline{\nabla}^\perp g_D)(\underline{x},\underline{y})\omega_t(\underline{y})d\underline{y} \tag{3.16}$$

Moreover

$$\lim_{\varepsilon\to 0} \sup_{t\in[0,T]} R(\omega_t^\varepsilon,\omega_t) = 0 \quad \text{for all} \quad T > 0 \tag{3.17}$$

Finally, the initial value problem (3.1) makes sense, has unique solution, and (3.2) holds.

Proof of Theorem 3.1. For notational simplicity we assume $\omega \geq 0$, $\int\omega(d\underline{x}) = 1$.

Let $\{\mu_t^i\}_{t\in[0,T]}$ $i = 1,2$ be two continuous, probability measure - valued functions. Consider the evolution problems in D:

$$\dot{\underline{x}}^i(t) = \underline{u}^{\varepsilon,i}(\underline{x}^i(t),t), \quad \underline{x}^i(t) \in D \tag{3.18}$$

where

$$\underline{u}^{\varepsilon,i}(\underline{x},t) = \int_D \underline{k}_\varepsilon(\underline{x},\underline{y})\mu_t^i(d\underline{y}) \tag{3.19}$$

The above problem is well posed because of the regularity of $\underline{k}_\varepsilon$ and the fact that $\underline{u}^{\varepsilon,i}\cdot\underline{n} = 0$ on ∂D.

We have

$$|\underline{k}_\varepsilon(\underline{x},\underline{y}) - \underline{k}_\varepsilon(\underline{x}',y)| \leq L_\varepsilon d(\underline{x},\underline{x}') \tag{3.20}$$

where

$$L_\varepsilon = \max(2\max|k_\varepsilon|, \tilde{L}_\varepsilon) \tag{3.21}$$

$\tilde{L}_\varepsilon$ being the Lipschitz constant of k_ε.

Thus, if $\underline{x}^i(t) \equiv \underline{x}^i(t,\underline{x})$ solves (3.18) with initial datum $\underline{x}$,

$$d(\underline{x}^1(t),\underline{x}^2(t)) \leq \int_0^t ds|\underline{u}^{\varepsilon,1}(\underline{x}^1(s),s) - \underline{u}^{\varepsilon,2}(\underline{x}^2(s),s)|$$

$$\leq \int_0^t ds\left|\int k_\varepsilon(\underline{x}^2(x) - \underline{y})(\mu_s^1(d\underline{y}) - \mu_s^2(d\underline{y}))\right| + L_\varepsilon\int_0^t ds\, d(\underline{x}^1(s),\underline{x}^2(s))$$

$$\leq \int_0^t ds\int \hat{P}_s(d\underline{y}_1,d\underline{y}_2)|k_\varepsilon(\underline{x}^2(s) - \underline{y}_1) - k_\varepsilon(\underline{x}^2(s) - \underline{y}_2)|$$

$$+ L_\varepsilon\int_0^t ds\, d(\underline{x}^1(s),\underline{x}^2(s)) \tag{3.22}$$

where $\hat{P}_s \in C(\mu_s^1,\mu_s^2)$. Applying (3.21), minimizing on $\hat{P}_s$ and using Gronwall lemma, we obtain

$$\sup_{0\leq t\leq T} d(\underline{x}^1(t),\underline{x}^2(t)) \leq [L_\varepsilon(\exp L_\varepsilon T)]\int_0^T ds\, R(\mu_s^1,\mu_s^2) \tag{3.23}$$

Notice now that the integral

$$\int \omega(d\underline{x}) f(\underline{x}^1(t,\underline{x}),\underline{x}^2(t,\underline{x})) \tag{3.24}$$

defines a joint representation of $\omega_t^1(d\underline{x})$ and $\omega_t^2(d\underline{x})$ solutions of the following *linear* equations:

$$\begin{cases} \frac{d}{dt}\omega_t^i(f) = \omega_t^i(\underline{u}^{\varepsilon,i}\cdot\nabla f) \\ \omega_0^i = \omega \end{cases} \tag{3.25}$$

We remark that ω_t^i is a probability measure because $\underline{\nabla}\cdot\underline{u}^{\varepsilon,i} = 0$. Integrating (3.23) in $\omega(d\underline{x})$ we obtain finally

$$\sup_{t\in[0,T]} R(\omega_t^1,\omega_t^2) \leq [L_\varepsilon(\exp L_\varepsilon T)]\int_0^T ds\, R(\mu_s^1,\mu_s^2). \tag{3.26}$$

Thus the map $\mu_t^1 \to \omega_t^1$ is a contraction in $C([0,T],M(1))$ for T small. The rest of the proof is straightforward. □

Proof of Theorem 3.2. Notice that, due to the conservation of the Lebesgue measure since $\underline{\nabla}\cdot\underline{u} = 0$:

$$\|\omega_t^\varepsilon\|_\infty = \|\omega\|_\infty\ , \qquad \|\omega_t^\varepsilon\|_1 = \|\omega\|_1\ . \tag{3.27}$$

Moreover it can be proved (see Appendix B at the end of this Section):

$$\int \omega(\underline{x})\,|\underline{k}_\varepsilon(\underline{y},\underline{x})|\,d\underline{x} \leq C(\|\omega\|_\infty + \|\omega\|_1) \tag{3.28$_1$}$$

$$\int \omega(\underline{x})\,|k_\varepsilon(\underline{y},\underline{x}) - k_\varepsilon(\underline{y}',\underline{x})|\,d\underline{x} \leq C(\|\omega\|_\infty + \|\omega\|_1)\varphi(\underline{y},\underline{y}') \qquad (3.28)_2$$

where

$$\varphi(\underline{x},\underline{x}') = \tilde{\varphi}(|\underline{x}-\underline{x}'|)$$
$$\tilde{\varphi}(r) = \begin{cases} r(1-\ln r) & \text{if} \quad 0<r<1 \\ 1 & \text{if} \quad r\geq 1 \end{cases} \qquad (3.29)$$

Here, and from now on, C denotes any positive constant. We want to proof that ω^ε is a Cauchy sequence for $\varepsilon \to 0$. To do this we compare $V^\varepsilon_{t,o}\underline{x}$ with $V^{\varepsilon'}_{t,o}\underline{x}$, $\varepsilon > \varepsilon'$.

We have, using (3.28) and equations of motion,

$$d(V^\varepsilon_{t,o}\underline{x},\, V^{\varepsilon'}_{t,o}\underline{x}) \leq tC\|\omega\|_\infty\varepsilon + C(\|\omega\|_1 + \|\omega\|_\infty) \int_0^t ds\{\varphi(V^\varepsilon_{s,o}\underline{x},\, V^{\varepsilon'}_{s,o}\underline{x}) + \left|\int d\underline{y}(\omega^\varepsilon_s(\underline{y}) - \omega^{\varepsilon'}_s(\underline{y}))\underline{k}_\varepsilon(V^{\varepsilon'}_{s,o}\underline{x},\underline{y})\right|\} \qquad (3.30)$$

We define

$$Y_1(\varepsilon,\varepsilon',t) = \int d\underline{x}\omega(\underline{x})\, d(V^\varepsilon_{t,o}\underline{x}, V^{\varepsilon'}_{t,o}\underline{x})$$
$$Y_2(\varepsilon,\varepsilon',t) = \int d\underline{x}\omega(\underline{x})\, \varphi(V^\varepsilon_{t,o}\underline{x}, V^{\varepsilon'}_{t,o}\underline{x}) \,. \qquad (3.31)$$

By the invariance of the Lebesgue measure for the flow $V^\varepsilon_{t,s}$ the last term in r.h.s. of (3.30) can be written as

$$\left|\int d\underline{y}\,\omega(\underline{y})k_\varepsilon(V^{\varepsilon'}_{s,o}\underline{x},V^{\varepsilon}_{s,o}\underline{y}) - k_\varepsilon(V^{\varepsilon'}_{s,o}\underline{x},V^{\varepsilon'}_{s,o}\underline{y})\right| \le C(\|\omega\|_1+\|\omega\|_\infty)\int d\underline{y}\,\omega(\underline{y})\,\varphi(V^{\varepsilon}_{s,o}\underline{y},V^{\varepsilon'}_{s,o}\underline{y}) \tag{3.32}$$

Hence, integrating on $\omega(\underline{x})d\underline{x}$ the estimate (3.30) we obtain

$$Y_1(\varepsilon,\varepsilon',t) \le C\|\omega\|_\infty t\varepsilon + C(\|\omega\|_\infty+\|\omega\|_1)\int_0^t ds\,Y_2(\varepsilon,\varepsilon',s) \tag{3.33}$$

By convexity inequality

$$Y_2(\varepsilon,\varepsilon',t) \le \tilde{\varphi}(Y_1(\varepsilon,\varepsilon',t)). \tag{3.34}$$

Finally consider the differential problem:

$$\begin{cases} \dot{Z} = C\tilde{\varphi}(r) \\ Z(0) = Z_o > 0. \end{cases} \tag{3.35}$$

It has the global solution

$$\begin{cases} Z(t,Z_o) = Z_o^{\exp-Ct}\exp\{1-e^{-Ct}\} & \text{if } Z < 1 \\ \qquad\qquad = 1 + C(t-t_o) & \text{if } Z \ge 1 \\ \text{where } t_o = \inf_t\{z(t) > 1\}, \quad \text{if } Z_o < 1. & \\ \text{and} & \\ \qquad Z(t,z_o) = Z_o + ct & \\ \text{if } Z_o > 1. & \end{cases} \tag{3.36}$$

Inserting (3.34) in (3.33) and making use of (3.36) we conclude that $Y_1(\varepsilon,\varepsilon',t)$ goes to zero as $\varepsilon \to 0$.

It is only matter of straightforward considerations to verify that the limit measure ω_t has $L_1 \cap L_\infty$ density, and uniquely satisfy the initial value problem (3.15), (3.16). □

We conclude this Section by discussing some regularity properties of the Euler flow in $\mathbb{R}^2$ that will be useful in the sequel. More precisely we want to prove that ω_t is spatially differentiable k-times if $\omega \in C^k$. Moreover we shall obtain also fixed time continuity properties of the Euler trajectories.

Suppose $\omega \in (L_1 \cap L_\infty)(\mathbb{R}^2)$, differentiable and such that $\|\nabla\omega\|_\infty \leq +\infty$. As we have seen

$$\omega_t(\underline{x}) = \omega(\underline{x}(-t)) \tag{3.37}$$

where

$$\underline{x}(t) = \underline{x} + \int_0^t \underline{u}(\underline{x}(s),s)ds. \tag{3.38}$$

By (3.28) we have for some positive constant C_1 depending only on ω:

$$|\underline{x}_1(t)-\underline{x}_2(t)| \leq |\underline{x}_1-\underline{x}_2| + C_1\int_0^t ds\, \varphi\,(\underline{x}_1(s),\underline{x}_2(s)) \tag{3.39}$$

implying (see (3.34), (3.36))

$$|\underline{x}_1(t) - \underline{x}_2(t)| \leq C_2|\underline{x}_1 - \underline{x}_2|^\alpha \tag{3.40}$$

(assuming $|x_1 - x_2| < 1$) where $0 < \alpha < 1$ and C_2 depend only on ω and some fixed time $T > 0$. Hence:

$$|\omega_t(\underline{x}_1) - \omega_t(\underline{x}_2)| \leq \|\nabla\omega\|_\infty |\underline{x}_1(-t) - \underline{x}_2(-t)| \leq \|\nabla\omega\|_\infty |\underline{x}_1 - \underline{x}_2|^\alpha \tag{3.41}$$

Since ω_t is Hölder continuous of exponent α, it can be proved that $\|\frac{\partial}{\partial x_\gamma} \underline{u}(\cdot,t)\|_\infty \leq C_3 < +\infty$, $\gamma = 1,2$ (see Appendix B). This is enough to improve estimate (3.39) to obtain

$$|\underline{x}_1(t) - \underline{x}_2(t)| \leq |\underline{x}_1 - \underline{x}_2| + C_3 \int_0^t ds\, |\underline{x}_1(s) - \underline{x}_2(s)| \tag{3.42}$$

implying, by time reversal

$$e^{-C_3 t} |\underline{x}_1 - \underline{x}_2| \leq |\underline{x}_1(t) - \underline{x}_2(t)| \leq e^{C_3 t} |\underline{x}_1 - \underline{x}_2| \; . \tag{3.43}$$

Furthermore, denoting by $\frac{\partial \underline{x}(t)}{\partial x_\alpha}$ $\alpha = 1,2$ the solution of the following integral equation

$$\frac{\partial x_\beta(t)}{\partial x_\alpha} = \frac{\partial x_\beta}{\partial x_\alpha} + \int_0^t \sum_{\gamma=1}^{2} \frac{\partial u_\beta}{\partial x_\gamma}(\underline{x}(s),s)\, \frac{\partial x_\gamma(s)}{\partial x_\alpha} \; , \tag{3.44}$$

one easily realizes that

$$\lim_{h\to 0} \frac{\underline{x}(t,\underline{x}+\underline{h}) - \underline{x}(t,\underline{x})}{h} = \frac{\partial \underline{x}(t)}{\partial x_\alpha} \tag{3.45}$$

where $h_\alpha = h$, $h_\beta = 0$, $\beta \neq \alpha$.

Hence

$$\frac{\partial \omega_t}{\partial x_\alpha}(\underline{x}) = \sum_\beta \frac{\partial \omega}{\partial x_\beta}(\underline{x}(-t)) \cdot \frac{\partial x_\beta(-t)}{\partial x_\alpha} \tag{3.46}$$

exists and the following bounds

$$\sup_{\alpha}\left\|\frac{\partial\omega_t}{\partial x_\alpha}\right\|_\infty \leq \sup_{\alpha}\left\|\frac{\partial\omega}{\partial x_\alpha}\right\|_\infty 2e^{C_3 t}$$

$$\sup_{\alpha}\left\|\frac{\partial\omega_t}{\partial x_\alpha}\right\|_1 \leq \sup_{\alpha}\left\|\frac{\partial\omega}{\partial x_\alpha}\right\|_1 2e^{C_3 t} \tag{3.47}$$

hold (the last in virtue of the invariance of the Lebesgue measure).

The procedure can be iterated to obtain the existence of higher derivatives of ω_t, assuming the existence of such derivatives at time zero.

Notes

There are many existence and uniqueness results for the Euler equation. One of the most satisfactory is due to Kato [Ka.1].

Our analysis follows [Ma.1] which contains all the details of the proof only sketched here.

The K.R. distance has been introduced by Kantorivich and Rubistein [K.1] and Vaserstein [V.1] for problems of transmission of information.

The idea of using K.R. distance for this kind of evolution problems is due to Dobrushin [Do.1]. Some basic fact on the K.R. distance and applications in Statistical Mechanics are in [Do.2].

Appendix A

Assume, without sake of generality a = 1 and

$$\mu_k(dx) = \frac{1}{N} \sum_{i=1}^{N} \delta_{x_i^k}(dx) \tag{A.1}$$

$k = 1,2$, $\{x_i^k\}_{i=1}^N \subset M$, be two Borel probability measures on M. Consider the N^2 points $y_{ij} \equiv (x_i^1, x_j^2)$ in $M \times M$. Then, any $\hat{P} \in C(\mu_1,\mu_2)$ takes the form:

$$\hat{P}(dy) = \sum_{i,j=1}^{N} a_{i,j}\delta_{y_{i,j}}(dy) , \quad y \in M \times M \tag{A.2}$$

where

$$\begin{gathered} a_{ij} \geq 0 \\ \sum_{j=1}^{N} a_{i,j} = \sum_{i=1}^{N} a_{i,j} = \frac{1}{N} . \end{gathered} \tag{A.3}$$

The elements $a_{i,j}$ form a matrix A

$$A = \begin{pmatrix} a_{11} & a_{12} & \cdots\cdots \\ a_{21} & & \\ \vdots & & \\ \vdots & & \end{pmatrix} \tag{A.4}$$

of positive elements for which the sum of any row or any column is $\frac{1}{N}$. Denote by A the set of such matrices. Then $C(\mu_1,\mu_2)$ is in one to one correspondence with the set A. The quantity $\int \hat{P}$ dx,dy)d(x,y) as function of $\hat{P}$ corresponds to the hyperplane

$$\sum_{i,j} a_{ij} d(x_i^1, x_j^2) \tag{A.5}$$

to be minimized on A, that is a convex compact set in $\mathbb{R}^{N^2}$. Hence the proof is achieved once one proves that the N! points of A defined as

$$a_{ij} = \begin{cases} \frac{1}{N} & \text{if } j = \pi(i) \\ 0 & \text{otherwise,} \end{cases} \tag{A.6}$$

are the only extremal points of A.

Denote by $\mathcal{D}$ the finite set defined in (A.6). Certainly $\mathcal{D}$ consists of extremal points. Let $A = \{a_{ij}\}_{i,j=1}^{N} \notin \mathcal{D}$ be an extremal point of A. Then for some i_1 there exists j_1 and j_2 s.t. $0 < a_{i_1j_1} < \frac{1}{N}$, $0 < a_{i_1j_2} < \frac{1}{N}$ $j_1 \neq j_2$. Moreover there exists i_2 s.t. $0 < a_{i_2j_2} < \frac{1}{N}$ and so on. Iterating the procedure, one can construct a graph in the matrix

$$\begin{array}{ccc} a_{i_1j_1} \longrightarrow & a_{i_1j_2} & \\ & \big\downarrow & \vdots \\ & a_{i_2j_2} \longrightarrow & a_{i_2j_3} \\ & & \vdots \end{array} \tag{A.7}$$

One easily realizes that a graph containing a closed loop $L = (a_{i_kj_k}\ a_{i_kj_{k+1}} \ldots\ldots a_{i_{n-1},j_n})$ with positive different elements, can be constructed.

Defining

$$A_{\pm} = \{a^{\pm}_{ij}\}$$

$$a^{\pm}_{ij} = a_{ij} \qquad \text{if } a_{ij} \notin L$$

$$a^{\pm}_{ij} = a_{i_s,j_{s+1}} \pm \varepsilon \ \text{ if } a_{i_s,j_{s+1}} \in L \tag{A.8}$$

$$= a_{i_s,j_s} \pm \varepsilon \ \text{ if } a_{i_s,j_s} \in L$$

$$\varepsilon < \min_{a_{ij} \in L} a_{ij} ,$$

it is easily verified that $A_{\pm} \in A$ because $a^{\pm}_{i_s,j_{s+1}} + a^{\pm}_{i_s,j_s} =$ $= a_{i_s,j_{s+1}} + a_{i_s,j_s}$ and $a^{\pm}_{i_s,j_s} + a^{\pm}_{i_{s+1},j_s} = a_{i_s,j_s} + a_{i_{s+1},j_s}$.

Finally

$$A = \frac{1}{2}(A^{+} + A^{-})$$

that contraddicts the hypothesis of extremality.

Appendix B

In this Appendix we want to prove estimates (3.28).

$$\int \omega(\underline{x})\,|\underline{k}_\varepsilon(\underline{y},\underline{x})|\,d\underline{x} \le C\int \frac{|\omega(\underline{x})|}{|\underline{y}-\underline{x}|}d\underline{x}$$

$$\le C\left\{\int_{|\underline{x}-\underline{y}|\le 1} \frac{|\omega(\underline{x})|}{|\underline{y}-\underline{x}|}d\underline{x} + \int_{|x-y|\ge 1} \cdots\cdots\right\}$$

$$\le C(\|\omega\|_\infty + \|\omega\|_1) \qquad \text{(B.1)}$$

Let $r = |\underline{y}-\underline{y}'| < 1$. (If $r \ge 1$ then $(3.28)_2$ is consequence of $(3.28)_1$). Consider $A = \{\underline{x}\,|\,|\underline{y}-\underline{x}| \le 2r\}$ then

$$\text{l.h.s. of } (3.28)_2 \le \int_A + \int_{A^C} (|\omega(\underline{x})|\,|k_\varepsilon(\underline{y},\underline{x}) - k_\varepsilon(\underline{y}',\underline{x})|) \qquad \text{(B.2)}$$

$$\int_A \cdots d\underline{x} \le C\|\omega\|_\infty\left(\int_A \frac{d\underline{x}}{|\underline{y}-\underline{x}|} + \int_A \frac{d\underline{x}}{|\underline{y}'-\underline{x}|}\right)$$

$$\le C\|\omega\|_\infty\left(\int_{|\underline{y}-\underline{x}|\le 2r} \frac{d\underline{x}}{|\underline{y}-\underline{x}|} + \int_{|\underline{x}-\underline{y}'|\le 3r} \frac{d\underline{x}}{|\underline{y}'-\underline{x}|}\right)$$

$$\le C\|\omega\|_\infty r \qquad \text{(B.3)}$$

Noticing that for all $\underline{y}''$ in the segment $\underline{y},\underline{y}'$ we have $|\underline{y}''-\underline{y}| > \frac{1}{2}|\underline{x}-\underline{y}|$, by the mean value theorem

$$\int_{A^C} \dots d\underline{x} \leq Cr\int_{A^C} \frac{|\omega(\underline{x})|\,d\underline{x}}{|\underline{y}-\underline{x}|^2}$$

$$\leq Cr\left(\int_{2r\leq|\underline{y}-\underline{x}|<2} \dots d\underline{x} + \int_{|\underline{y}-\underline{x}|>2} \dots d\underline{x}\right)$$

$$\leq Cr\|\omega\|_\infty \int_{2r}^{2} \frac{d\rho}{\rho} + Cr\|\omega\|_1 \qquad \text{(B.4)}$$

that concludes the proof of $(3.28)_2$.

We now prove the estimate $\|\frac{\partial}{\partial x_\gamma}\underline{u}(\cdot,t)\|_\infty \leq$ Const. assuming $\omega \in L_1 \cap L_\infty(\mathbb{R}^2)$ and Hölder continuous of exponent α. Since we know under this hypothesis that $\underline{u} = -\underline{\nabla}^\perp\Delta^{-1}\omega$ is differentiable, it is enough to prove the estimate

$$|\underline{u}(\underline{y}) - \underline{u}(\underline{y}')| \leq \text{Const.}\,|\underline{y} - \underline{y}'| \qquad \text{(B.5)}$$

Putting as above $|\underline{y}-\underline{y}'| = r$, we limit ourselves to the case $r < 1$.

Let Σ_R be a circle of radius R around the point $\underline{y}$.

Fixed $R > 2$:

$$\left|\int_{\Sigma_R^C} \omega(\underline{x})\underline{k}(\underline{y}-\underline{x}) - k(\underline{y}'-\underline{x})\right| \leq Cr\|\omega\|_1 \qquad \text{(B.6)}$$

On the other hand, by Gauss theorem

$$|\omega(\underline{y})\int_{\Sigma_R} (k(\underline{y}-\underline{x}) - k(\underline{y}'-\underline{x}))d\underline{x}| \leq |\omega(\underline{y})|\,|\int_{\Sigma_R} \underline{k}(\underline{y}'-\underline{x})d\underline{x}|$$

$$\leq \text{Const.}\int_{R-r}^{R+r} d\rho = \text{Const.}\, r \qquad \text{(B.7)}$$

Hence

$$|\underline{u}(\underline{y})-\underline{u}(\underline{y}')| \leq |\int_{\Sigma_R} (\omega(\underline{x})-\omega(\underline{y}))(k(\underline{y}-\underline{x})-k(\underline{y}'-\underline{x}))d\underline{x}|$$

$$|\int_{\complement\Sigma_R} \omega(\underline{x})[k(\underline{y}-\underline{x})-k(\underline{y}'-\underline{x})]d\underline{x}| + |\int_{\Sigma_R} \omega(y)(k(\underline{y}-\underline{x})-k(\underline{y}'-\underline{x}))d\underline{x}| \ . \tag{B.8}$$

and it remains to estimate the first integral in the r.h.s. of (B.8).

Proceeding as in proving $(3.28)_2$ (replacing $\omega(\underline{x})$ by $\omega(\underline{x}) - \omega(\underline{y})$) it is enough to prove

$$\int_{2r\leq|\underline{y}-\underline{x}|<2} \frac{|\omega(\underline{x})-\omega(\underline{y})|}{|\underline{x}-\underline{y}|^2}\, d\underline{x} \leq \text{Const.} \tag{B.9}$$

Estimate (B.9) holds in virtue of the Hölder continuity of ω and hence the proof of (B.5) is achieved.

4. FURTHER CONSIDERATIONS ON VORTEX MODEL

In this Section we deal, mostly, with the problem of justifying the vortex model in a more convincing way than we did in Section 2. Our aim is to deduce the vortex equation starting from the Euler equations, assumed as the basic equations for the evolution of an ideal fluid. To do this, a reasonable approach is the following. Suppose to have a vorticity distribution sharply concentrated in N disjoint blobs of intensities a_i around the points $\underline{x}_i$, and sufficiently smooth to be evolved by means of the Euler dynamics. If we prove that, the vorticity distribution is, at time t, similar to $\sum_{i=1}^{N} a_i \delta_{\underline{x}_i(t)}$, where $\{\underline{x}_i(t)\}_{i=1}^{N}$ is the solution of the vortex dynamics with initial data $\{\underline{x}_i\}_{i=1}^{N}$, then this means that the selfinteraction is actually negligible so giving a satisfactory dynamical justification of the vortex model.

In the sequel we perform this program for short times. We first consider the case $D = \mathbb{R}^2$.

We begin by establishing some properties of the motion of a single blob in a regular external velocity field.

Theorem 4.1. Let us define, for an open bounded region Λ^ε, such that meas $\Lambda^\varepsilon = \pi\varepsilon^2$,

$$\omega^\varepsilon(\underline{x}) = \frac{1}{\pi\varepsilon^2} \chi_{\Lambda^\varepsilon}(\underline{x}), \tag{4.1}$$

where $\chi_{\Lambda^\varepsilon}$ denotes the characteristic function of the set Λ^ε.

Suppose:

$$\lim_{\varepsilon \to 0} \omega^{\varepsilon}(f) = f(\bar{\underline{x}}) , \quad \bar{\underline{x}} \in \mathbb{R}^2 , \tag{4.2}$$

for any continuous founded f.

Let $\underline{F}$ be a divergence free uniformly bounded vector field satisfying the Lipschitz condition

$$|\underline{F}(\underline{x},t) - \underline{F}(\underline{x}',t')| \leq k\{|\underline{x}-\underline{x}'| + |t-t'|\}, \quad k \in \mathbb{R}^+. \tag{4.3}$$

and $\omega_t^{\varepsilon}(\underline{x})$ be the solution of the Euler equations with external field $\underline{F}$, i.e.

$$\begin{cases} \frac{d}{dt} \omega_t^{\varepsilon}(f) = \omega_t^{\varepsilon}(\underline{u}_t^{\varepsilon} \cdot \underline{\nabla} f) + \omega_t^{\varepsilon}(\underline{F} \cdot \underline{\nabla} f) \\ \omega_o^{\varepsilon} = \omega^{\varepsilon} , \quad \underline{u}_t^{\varepsilon} = -\underline{\nabla}^{\perp} \Delta^{-1} \omega^{\varepsilon} . \end{cases} \tag{4.4}$$

Then

$$\lim_{\varepsilon \to 0} \omega_t^{\varepsilon}(f) = f(\underline{B}(t)) \tag{4.5}$$

where $\underline{B}(t)$ solves

$$\begin{cases} \dot{\underline{B}}(t) = \underline{F}(\underline{B}(t),t) \\ \underline{B}(0) = \bar{\underline{x}} . \end{cases} \tag{4.6}$$

Proof. The difficulty in proving Theorem 4.1, lies, obviously, in the singular character of the interaction kernel. In fact if one looks at the paths of two particles of the blob, they are very different in the limit $\varepsilon \to 0$. Actually, the lenght of the characteristics of points starting from the blob, diverge in the same limit. To overcome this difficulty we investigate the motion of the center of vorticity, that does not follow a particle path for $\varepsilon > 0$.

Let us define

$$\underline{B}_\varepsilon(t) = \int d\underline{x}\,\underline{x}\,\omega_t^\varepsilon(\underline{x}) \quad . \tag{4.7}$$

Then

$$\dot{\underline{B}}_\varepsilon(t) = \int d\underline{x}\,\underline{F}(\underline{x},t)\omega_t^\varepsilon(\underline{x}) \tag{4.8}$$

after using the identiy

$$\int_{\Lambda_t^\varepsilon} d\underline{x} \int_{\Lambda_t^\varepsilon} d\underline{y}\,k(\underline{x}-\underline{y}) = 0 \quad . \tag{4.9}$$

Thus we have to prove that Λ_t^ε, the evolution of the region Λ^ε along the characteristics of the Euler flow, is in a small neighborhood of the point $\underline{B}_\varepsilon(t)$. To control this we introduce the second moment

$$I_\varepsilon(t) = \int (\underline{x}-\underline{B}_\varepsilon(t))^2\omega_t^\varepsilon(\underline{x})d\underline{x} \quad . \tag{4.10}$$

Also I_ε, as $\underline{B}_\varepsilon$, would be a conserved quantity if $\underline{F} \equiv 0$.

We have, by direct computation:

$$\dot{I}_\varepsilon(t) = 2\int (\underline{x}-\underline{B}_\varepsilon(t))\cdot\underline{F}(\underline{x},t)\omega_t^\varepsilon(\underline{x})d\underline{x} \tag{4.11}$$

after using

$$\int_{\Lambda_t^\varepsilon} d\underline{x} \int_{\Lambda_t^\varepsilon} d\underline{y}\,\underline{x}\cdot\underline{k}(\underline{x}-\underline{y}) = \frac{1}{2}\int_{\Lambda_t^\varepsilon} d\underline{x} \int_{\Lambda_t^\varepsilon} d\underline{y}(\underline{x}-\underline{y})\cdot\underline{k}(\underline{x}-\underline{y}) = 0 \tag{4.12}$$

By (4.11)

$$|\dot{I}_\varepsilon(t)| \leq 2k\int (\underline{x}-\underline{B}_\varepsilon(t))^2 \omega_t^\varepsilon(\underline{x})\,d\underline{x} \tag{4.13}$$

and hence

$$I_\varepsilon(t) \leq I_\varepsilon(0)e^{2kt}\ . \tag{4.14}$$

The proof of Theorem 4.1 is then easily achieved. □

Now we want to prove that the blob of vorticity remains in a circle of fixed radius uniformly in ε, at least for short time. This is the content of the following theorem that will be, combined with Theorem 4.1, the main ingredient in proving the desired convergence result.

Theorem 4.2. Suppose the hypotheses of Theorem 4.1 hold, and in addition

$$\Lambda_t^\varepsilon \subset \textstyle\sum_{\alpha\varepsilon}(\bar{\underline{x}}), \qquad \alpha > 0, \tag{4.15}$$

where $\sum_R(\underline{x})$ is the circle of radius R, centered at $\underline{x}$.

Then, for all $R > \alpha\varepsilon_0$, there exists $t^* > 0$ such that, for sufficiently small $\bar{\varepsilon} < \varepsilon_0$,

$$\bigcup_{\varepsilon\leq\bar{\varepsilon}} \Lambda_t^\varepsilon \subset \textstyle\sum_R(\underline{B}(t)) \tag{4.16}$$

for $0 \leq t < t^*$.

Proof. Let us define:

$$A_\varepsilon(t) = \Lambda_t^\varepsilon \cap R^2/\textstyle\sum_d(\underline{B}_\varepsilon(t)). \tag{4.17}$$

We have by (4.15) and (4.14)

$$\frac{d^2}{\pi\varepsilon^2}\,\mathrm{meas}A_\varepsilon(t) \le I_\varepsilon(t) \le C_1\varepsilon^2 e^{2kt} \tag{4.18}$$

for $C_1 > 0$. Hence the velocity field generated by the vorticity supported in $A_\varepsilon(t)$ is bounded by

$$\frac{1}{\pi\varepsilon^2}\sup_{\underline{x}} \int_{A_i^\varepsilon(t)} \frac{1}{|\underline{x}-\underline{y}|}d\underline{y} \le \frac{1}{\pi\varepsilon^2}\int_{\sum_\gamma(0)} \frac{1}{|\underline{y}|}d\underline{y} \tag{4.19}$$

where

$$\gamma = \sqrt{\frac{\mathrm{meas}A_\varepsilon(t)}{\pi}}\ .$$

Therefore, the r.h.s. of the inequality (4.19) is bounded by:

$$\frac{C_2}{d}\,e^{2kt}\ . \tag{4.20}$$

and a point, starting on Λ^ε at time zero, and outside $\sum_{2d}(\underline{B}_\varepsilon(t))$, has a velocity bounded by

$$\frac{1}{2\pi d} + \frac{C_2}{d}e^{2kt} + \|\underline{F}\|_\infty \tag{4.21}$$

where the first contribution is a bound of the velocity produced by the vorticity inside $\sum_d(\underline{B}_\varepsilon(t))$, the second follows by (4.20) and the last one, is due to the regular field $\underline{F}$($\|\underline{F}\|_\infty$ = $= \sup\limits_{0<t\le T}\sup\limits_{x}|\underline{F}(x,t)|$).

Finally, denoting by $\underline{x}^\varepsilon(t)$ the characteristics of a point starting from Λ^ε at time zero:

$$|\underline{x}^{\varepsilon}(t) - \underline{B}(t)| \leq |\underline{x}^{\varepsilon}(t) - \underline{B}^{\varepsilon}(t)| + |\underline{B}^{\varepsilon}(t) - \underline{B}(t)|$$

$$\leq \alpha\varepsilon + 2d + \left(\frac{1}{2\pi d} + \frac{C_2}{d} e^{kt} + \|\underline{F}\|_{\infty}\right) t \tag{4.22}$$

assuming ε so small that $|B^{\varepsilon}(t) - B(t)| < d$. The thesis follows by choosing a sufficiently small d satisfying

$$\alpha\varepsilon + 2d < R \tag{4.23}$$

and consequently t so small that

$$\left(\frac{1}{2\pi d} + \frac{C_2}{d} e^{kt} + \|\underline{F}\|_{\infty}\right) t < (R - \alpha\varepsilon - 2d) \tag{4.24}$$

□

The connection between the vortex model and the Euler evolution is established by the following theorem.

Theorem 4.3. Let us define:

$$\omega^{\varepsilon}(\underline{x}) = \frac{1}{\pi\varepsilon^2} \sum_{i=1}^{N} a_i \chi_{\Lambda_i^{\varepsilon}}(\underline{x}) \tag{4.25}$$

where Λ_i^{ε} are disjoint open regions such that meas $\Lambda_i^{\varepsilon} = \pi\varepsilon^2$.

Suppose

$$\lim_{\varepsilon \to 0} \omega^{\varepsilon}(f) = \sum_{i=1}^{N} f(\underline{\tilde{x}}_i), \quad \underline{\tilde{x}}_i \in \mathbb{R}^2 \tag{4.26}$$

for bounded continuous f.

Moreover assume

$$\Lambda_i^{\varepsilon} \subset \Sigma_{\alpha_i \varepsilon}(\underline{\tilde{x}}_i), \quad \alpha > 0 . \tag{4.27}$$

Then, denoting by ω_t^ε the weak solution of the Euler equation with initial datum ω^ε, there exists a t* for which:

$$\lim_{\varepsilon\to 0} \omega_t^\varepsilon(f) = \sum_{i=1}^{N} a_i f(\underline{x}_i(t)) \qquad (4.28)$$

for $t < t^*$, where $\underline{x}_i(t)$ denotes the solution of the vortex dynamics for the vortices $(a_i, \underline{x}_i(t))$ starting from $\tilde{\underline{x}}_i$ at time zero.

We do not give the proof of this theorem that is a rather easy consequence of Theorems 4.1 and 4.2. In any case the reader will realize that the proof goes along the same lines of the proof of Theorem 4.4 below.

Analogous results can be obtained in case of a simply connected domain with boundaries. In this case one has to take into account the interaction of the blobs with the boundaries in the approximate conservation laws for $\underline{B}_\varepsilon$ and I_ε. This gives rise to the extra term $\frac{1}{2} \underline{\nabla}_i^\perp a_i \tilde{\gamma}_D$ in eq. (2.9).

The techniques used in the first part of this Section, allow us to prove the following natural fact. Suppose to have a system of N vortices divided into n weakly interacting clusters each of them constituted by strongly interacting vortices of the same sign. Then one expects the motion of the system to be similar to the motion of the centers of vorticity of the clusters, each of them thought as single vortex equipped with the total vorticity of the associated cluster. This is a well known theorem of Classical Mechanics, the only difficulty of its proof lying on the singularity of the interaction.

Let us establish a more precise statement. Consider a family of systems of N vortices $\underline{x}_1^\varepsilon \ldots\ldots \underline{x}_N^\varepsilon$, $\varepsilon \in (0,1)$ with positive intensities $a_1 \ldots\ldots a_N$. We divide the vortices into n clusters according to a partition $J_1 \ldots J_n$ of $\{1 \ldots\ldots N\}$ and denote by

$$\underline{z}_k^\varepsilon = \frac{1}{A_k} \sum_{i \in J_k} a_i \underline{x}_i^\varepsilon$$
$$A_k = \sum_{i \in J_k} a_i \qquad k = 1 \ldots n \tag{4.29}$$

the centers of vorticity and the total charge of the clusters. We assume

$$\{\underline{x}_i^\varepsilon \mid i \in J_k\} \in \sum_\varepsilon (\underline{z}_k^\varepsilon) \tag{4.30}$$

and

$$\lim_{\varepsilon \to 0} \underline{z}_k^\varepsilon = \underline{z}_k \ . \tag{4.31}$$

We denote by $\{\underline{x}_i^\varepsilon(t)\}_{i=1}^N$ the vortex dynamics with initial condition $\{\underline{x}_i^\varepsilon\}_{i=1}^N$, by $\{\underline{z}_k^\varepsilon(t)\}_{k=1}^n$ the evolution of the centers of the clusters and by $\{z_k(t)\}_{k=1}^n$ the solution of the reduced vortex dynamics

$$\begin{cases} \dot{\underline{z}}_k(t) = \sum_{\substack{h=1 \\ h \neq k}}^{n} A_h \underline{k}(\underline{z}_k(t) - \underline{z}_h(t)) \\ \underline{z}_k(0) = \underline{z}_k \end{cases} \tag{4.32}$$

We have:

<u>Theorem 4.4.</u>

$$\lim_{\varepsilon \to 0} \underline{z}_k^\varepsilon(t) = \underline{z}_k(t) \ , \quad \text{for all} \quad t > 0, \quad k = 1 \ldots\ldots n \tag{4.33}$$

Proof. The idea is based on the previous considerations. First of all, for a single cluster $\{\underline{x}_1 \ldots \underline{x}_m\}$ moving in an external smooth field $\underline{F}$ we have (accordingly to Theorem 4.1)

$$\dot{\underline{z}}(t) = \underline{F}(\underline{z}(t),t) \tag{4.34}$$

$$|I(t)| \leq |I(0)|e^{2kt} \tag{4.35}$$

where $\underline{z}(t)$ is the center of vorticity of the cluster and

$$I(t) = \sum_{i=1}^{m} (\underline{x}_i(t) - \underline{z}(t))^2 a_i \, . \tag{4.36}$$

Let us introduce a regularized interaction $\underline{k}_\eta$ (η sufficiently small to be defined later) following definition (3.4) (in case $D = \mathbb{R}^2$) and define $\tilde{\underline{x}}_i^\varepsilon(t)$, $t > 0$, $i = 1 \ldots N$, the vortex dynamics in which two vortices of different clusters interact via $\underline{k}_\eta$, while vortices of the same cluster interact via $\underline{k}$

For an arbitrary, but fixed, $T > 0$, we choose η so small that

$$\min_{h \neq k} \inf_{0 \leq t \leq T} |\underline{z}_h(t) - \underline{z}_k(t)| > 10\eta \tag{4.37}$$

Finally, denoting by $\tilde{\underline{z}}_h^\varepsilon(t)$ the center of the cluster $\{\tilde{\underline{x}}_i\}_{i \in J_h}$, we have for $t < T$:

$$|\underset{\sim}{\tilde{z}}_h^{\varepsilon}(t)-\underset{\sim}{z}_h(t)| \le |\underset{\sim}{z}_h^{\varepsilon}-\underset{\sim}{z}_h|+$$

$$\int_0^t ds \sum_{\substack{k\neq h\\k=1}}^{n} \sum_{i\in J_k} a_i[\underset{\sim}{k}_\eta(\tilde{z}_h^{\varepsilon}(s)-\tilde{x}_i^{\varepsilon}(s))-\underline{k}(z_n(s)-z_k(s))]$$

$$\le |\underset{\sim}{z}_h^{\varepsilon}-\underset{\sim}{z}_h|+L_\eta\int_0^t ds \sum_{\substack{k\neq h\\k=1}}^{n} A_k\{|\underset{\sim}{\tilde{z}}_h^{\varepsilon}(s)-\underset{\sim}{z}_h(s)|+|\underset{\sim}{\tilde{z}}_k^{\varepsilon}(s)-\underset{\sim}{z}_k(s)|\}$$

$$+L_\eta\int_0^t ds \sum_{\substack{k\neq h\\k=1}}^{n} \sum_{i\in J_k} a_i|\underset{\sim}{\tilde{z}}_k^{\varepsilon}(s)-\underset{\sim}{\tilde{x}}_i^{\varepsilon}(s)| \qquad (4.38)$$

Where L_η is a positive constant depending on η.
The last integral in (4.38) is bounded by

$$L_\eta\int_0^t ds \sum_{\substack{k\neq h\\k=1}}^{n} [\sum_{i\in J_k} a_i(\underset{\sim}{\tilde{z}}_h^{\varepsilon}(s)-\underset{\sim}{\tilde{x}}_i^{\varepsilon}(t))^2]^{\frac{1}{2}}\sqrt{A_k} \qquad (4.39)$$

hence by (4.35), (4.30) and (4.37) we obtain

$$\lim_{\varepsilon\to 0} \underset{\sim}{\tilde{z}}_h^{\varepsilon}(t) = \underset{\sim}{z}_h(t), \quad 0 \le t \le T, \quad h=1\ldots n \qquad (4.40)$$

To complete the proof of the Theorem it is enough to prove that for sufficiently small ε

$$\underset{\sim}{\tilde{z}}_h^{\varepsilon}(t) = \underset{\sim}{z}_h^{\varepsilon}(t), \quad 0 \le t \le T, \quad h=1\ldots h \qquad (4.41)$$

This is again consequence of (4.35) implying that each cluster is contained, up to the time T, in $\sum_{\frac{\eta}{4}}(\underset{\sim}{z}_k(t))$, for sufficiently small ε.

□

Remark 1. The hypothesis that all vortices have the same sign cannot be avoided, in general. If a pair, of vortices of opposite intensity constitutes a cluster, they move with very high speed and, at least in principle, could strongly interact with other clusters.

Remark 2. We obtained in Theorem 4.4 a result that is valid for all times, while, the analogous result stated in Theorem 4.3 is proved for short times only. Actually, in Theorem 4.4, there is a minimal vorticity intensity (i.e. $\min_i a_i$) for which (4.35) is enough to ensure absence of vortices of the cluster outside a small circle around its center. On the contrary in the continuous case (Theorem 4.2), we cannot exclude a priori the presence of a small amount of vorticity outside any small circle, giving rise to finite perturbation of the velocity field.

Another interesting situation in which the vorticity is concentrated on manifolds of dimension less than two, is the so called vortex sheet.

We consider a smooth curve in the x,y plane, described by the function $y = \varphi(x)$. On such a curve is distributed a vorticity density given by the function $\gamma = \gamma(x)$. The vorticity distribution is then given by

$$\omega(x,y)dxdy = \gamma(x)\delta(y-\varphi(x))dxdy \ . \tag{4.42}$$

We want to deduce the motion of the pair $(\gamma,\varphi) \to (\gamma(\cdot,t),\varphi(\cdot,t))$ accordingly to the Euler evolution. In this case we expect less troubles than in case of point vorticity distribution, because the velocity field generated by a vortex sheet like (4.42) is bounded, so a renormalization of the interaction is unnecessary.

It is well known from potential theory [Ke.1], that the tangential component of the velocity suffers a jump crossing the curve $y = \varphi(x)$, while the normal component is continuous. We need to define the velocity on the sheet because it will be responsible of the motion of the sheet itself.

A reasonable choice for the tangent part would be $\frac{u_\tau^+ + u_\tau^-}{2}$ whre $u_\tau^\pm$ are the two limits from each side of the sheet of the tangent component of the velocity. With this definition the vector

$$\underline{v}(x) = \underline{u}(x,\varphi(x)), \tag{4.43}$$

where $\underline{u}(x,\varphi(x))$ is the symmetrized velocity on the sheet $(x,\varphi(x))$, has components

$$\begin{aligned} v^{(1)}(x) &= \frac{1}{2\pi}\int \frac{\varphi(x)-\varphi(x')}{(x-x')^2+(\varphi(x)-\varphi(x'))^2}\gamma(x')dx' \\ v^{(2)}(x) &= -\frac{1}{2\pi}\int \frac{x-x'}{(x-x')^2+(\varphi(x)-\varphi(x'))^2}\gamma(x')dx' \end{aligned} \tag{4.44}$$

where the above integrals have to be understood in the sense of Cauchy principal value.

Then we expect for φ and γ the following evolution equations

$$\begin{cases} \frac{\partial\varphi}{\partial t}(x,t) = -v^{(1)}(x,t)\frac{\partial\varphi}{\partial x}(x,t) + v^{(2)}(x,t) & (4.45)_1 \\ \frac{\partial\gamma}{\partial t}(x,t) + \frac{\partial}{\partial x}(v^{(1)}\gamma)(x,t) = 0 & (4.46)_2 \end{cases}$$

In fact $(4.45)_1$ follows by computing the infinitesimal displace ment of the point $(x,\varphi(x))$ under $\underline{v}$ and $(4.45)_2$ by the conserva tion of the vorticity, implying a continuity equation.

It is not hard to see that eq.s (4.45) solve, at least formally, the Euler eq.s in weak form.

In fact by taking a sufficiently smooth test function f, defining

$$\omega_t(x,y)dxdy = \gamma(x,t)\delta(y-\varphi(x,t))dxdy \qquad (4.46)$$

we have

$$\frac{d}{dt}\omega_t(f) = \int\frac{\partial}{\partial t}\gamma(x,t)f(x,\varphi(x,t))dx + \int\gamma(x,t)\frac{\partial f}{\partial y}(x,\varphi(x,t))\frac{\partial\varphi}{\partial t}(x,t)dx$$

$$= -\int\frac{\partial}{\partial x}(v^{(1)}\gamma)(x,t)f(x,\varphi(x,t))dx +$$

$$-\int\gamma(x,t)\frac{\partial f}{\partial y}(x,\varphi(x,t))\frac{\partial\varphi}{\partial x}(x,t)\cdot v^{(1)}(x,t)dx$$

$$+\int\gamma(x,t)\frac{\partial f}{\partial y}(x,\varphi(x,t))\cdot v^{(2)}(x,t)dx \qquad (4.47)$$

Integrating by parts the first integral in the r.h.s. of (4.47) we get

$$\int (v^{(1)}\gamma)(x,t)\frac{\partial f}{\partial x}(x,\varphi(x,t))dx + \int (v^{(1)}\gamma)(x,t)\frac{\partial f}{\partial y}(x,\varphi(x,t))\frac{\partial \varphi}{\partial x}(x,t) \tag{4.48}$$

and hence the Euler eq.s are satisfied.

At less formal level the situation is more intricate. The only result we know concerning the correct position of the problem (4.45) (valid also in three dimensional cases) states existence for all times and uniqueness and analiticity of the solutions for short times [Su.1].

Possible collapses for the system (4.45) are related to the so called Helmholtz-Kelvin instability. In fact if one linearize eq.s (4.45) around the stationary solution $\varphi \equiv 0$ $\gamma = 1$, (describing two opposite uniform flows in the upper and lower halfplane) the linear equation governing the motion of a disturbance, exhibits solutions growing exponentially in time. (See [Ba.1] and [Su.1] for a complete analytical treatment). For example a sinusoidal disturbance of uniform vorticity density has the tendency to accumulate vorticity at the nodes and therefore to roll up [Ba.1], [Bi.1], [Sa.1].

5. A MEAN FIELD LIMIT

In this Section we want to investigate another interesting connection between the vortex motion and the Euler flow.

The basic idea is, roughly, the following. The Euler flow may be thought as the motion of infinitely many vortices of infinitesimal intensity. If this is true, at least in some sense, one can hope to approximate the Euler flow by a finite dimensional motion of sufficiently many vortices. This is interesting both for a better understanding of the Euler flow and for computational purposes.

We consider a point particles system satisfying the following evolution equation:

$$\dot{\underline{x}}_i = \sum_{j=1}^{n} a_j \underline{k}_\varepsilon(\underline{x}_i, \underline{x}_j) \tag{5.1}$$

where k_ε has been defined in (3.4).

This is a regularized version of the vortex model in which the interaction of a single vortex with the boundary is neglected. Such hypothesis is not essential in what follows: we do it for notational simplicity. In case $D = \mathbb{R}^2$ the above model obviously coincides with the usual regularized vortex model.

Let

$$\mu^n(d\underline{x}) = \sum_{i=1}^{n} a_i \delta_{\underline{x}_i}(d\underline{x}), \quad \underline{x}_i \in D \tag{5.2}$$

be a measure in $M(a^+, a^-)$, where $a^\pm$ are the total charge of positive and negative vortices and

$$\mu_t^n(d\underline{x}) = \sum_{i=1}^{n} a_i \delta_{\underline{x}_i(t)}(d\underline{x}), \tag{5.3}$$

the evolved measure obtained by solving the initial value problem associated to (5.1), with initial datum $\underline{x}_1 \ldots \underline{x}_n$. We have, for all $f \in C^\infty(D)$

$$\begin{cases} \frac{d}{dt}\mu_t^n(f) = \mu_t^n(\underline{u}^\varepsilon \cdot \underline{\nabla} f) \\ \mu_o^n = \mu^n \end{cases} \tag{5.4}$$

where, as in (3.6), $\underline{u}^\varepsilon(\underline{x},t) = \int \underline{k}_\varepsilon(\underline{x},\underline{y})\mu_t^n(d\underline{y})$. Hence μ_t^n is a particular solution of the initial value problem (3.5), (3.6).

Theorem 5.1. Let the hypothesis of Theorem 3.2 be satisfied and $L_1 \cap L_\infty(D) \ni \omega \to \omega_t \quad t \in [0,T]$ be the solution of the weak form of the Euler equation. Suppose that

$$\lim_{n\to\infty} R(\omega,\mu^n) = 0 . \tag{5.5}$$

Then, for all sequences $\varepsilon = \varepsilon(n)$ such that

$$\lim_{n\to\infty} R(\omega,\mu^n)\exp[aL_\varepsilon(T + e^{aL_\varepsilon T})] = 0 \tag{5.6}$$

where

$$\begin{aligned} a &= a^+ + a^- , \\ L_\varepsilon &= \max(\tilde{L}_\varepsilon, 2\max k_\varepsilon), \\ \tilde{L}_\varepsilon &= \text{Lipschitz constant of } k_\varepsilon , \end{aligned} \tag{5.7}$$

we have:

$$\lim_{n\to\infty} \sup_{t\in[0,T]} R(\omega_t,\mu_t^n) = 0 \tag{5.8}$$

Proof. Let $\omega_{i,t}^{\varepsilon}$ be two solutions of the initial value problem (3.5), (3.6) with initial conditions ω_i, $i=1,2$.

Then:

$$\sup_{t\in[0,T]} R(\omega_{1,t},\omega_{2,t}) \leq R(\omega_1,\omega_2)\exp[L_\varepsilon(T+e^{L_\varepsilon T})] \tag{5.9}$$

In fact:

$$R(\omega_{1,t}^{\varepsilon},\omega_{2,t}^{\varepsilon}) \leq R(\omega_{1,t}^{\varepsilon},\bar{\omega}_{1,t}^{\varepsilon}) + R(\bar{\omega}_{1,t}^{\varepsilon},\omega_{2,t}^{\varepsilon}) \tag{5.10}$$

where $\bar{\omega}_{1,t}^{\varepsilon}$ solves the *linear* problem

$$\begin{cases} \dfrac{d}{dt}\bar{\omega}_{1,t}^{\varepsilon}(f) = \bar{\omega}_{1,t}^{\varepsilon}(\underline{u}^{2,\varepsilon}\cdot\underline{\nabla}f) \\ \bar{\omega}_{1,o}^{\varepsilon} = \omega_1 \end{cases} \tag{5.11}$$

and where

$$\underline{u}^{2,\varepsilon}(\underline{x},t) = \int \underline{k}_\varepsilon(\underline{x},\underline{y})\omega_{2,t}^{\varepsilon}(d\underline{y}) \;. \tag{5.12}$$

It can be proved, rather easily, the following property of continuity w.r.t. initial conditions:

$$\sup_{t\in[0,T]} R(\bar{\omega}_{1,t}^{\varepsilon},\omega_{2,t}^{\varepsilon}) \leq R(\omega_1,\omega_2)\exp L_\varepsilon T \tag{5.13}$$

On the other hand we have, by (3.26)

$$\sup_{s\leq t} R(\omega_{1,s}^{\varepsilon},\bar{\omega}_{1,s}^{\varepsilon}) \leq L_\varepsilon(\exp L_\varepsilon T)\int_0^t ds \sup_{r\leq s} R(\omega_{1,r}^{\varepsilon},\omega_{2,r}^{\varepsilon}) \tag{5.14}$$

and hence

$$\sup_{s\leq t} R(\omega^{\varepsilon}_{1,s},\omega^{\varepsilon}_{2,s}) \leq R(\omega_1,\omega_2)\exp L_{\varepsilon}T +$$

$$+ L_{\varepsilon}(\exp L_{\varepsilon}T)\int_0^t ds \sup_{r\leq s} R(\omega^{\varepsilon}_{1,r},\omega^{\varepsilon}_{2,r}) \tag{5.15}$$

implying (5.9) by the use of Grouwall's lemma.

For a given sequence $\varepsilon = \varepsilon(n)$, let us denote $\omega^n_t = \omega^{\varepsilon(n)}_t$. Then:

$$R(\omega_t,\mu^n_t) \leq R(\omega_t,\omega^n_t) + R(\omega^n_t,\mu^n_t) \tag{5.16}$$

$$R(\omega^n_t,\mu^n_t) \leq R(\omega,\mu^n)\exp[L_{\varepsilon}(T+e^{L_{\varepsilon}T})] \tag{5.17}$$

by (5.9).

Since $\sup_{t\in[0,T]} R(\omega_t,\omega^n_t) \to 0$ (see Th. 3.2) the thesis follows by (5.17) and (5.6). □

Theorem 4.1 works under very general assumptions. Nevertheless it is not very satisfactory from a quantitative point of view. The choice of the cutoff lenght $\varepsilon = \varepsilon(n)$ seems in fact very problematic. From one side ε has to converge to zero fastly to make $R(\omega_t,\omega^h_t) \approx \varepsilon$ (see (3.33) and (3.35)) small. On the other side ε has to converge slowly to satisfy the orrible condition (5.6).

Recent papers (see Notes at the end of the Section) give a much better fixed time approximation under more restrictive assumptions on the initial vorticity profile ω and assuming $D = \mathbb{R}^2$.

We try to explain some ideas of these papers following [Bea.1].

From now on in this Section we assume $L_\varepsilon \leq \text{const}.\varepsilon^{-2}$ as follows by the choice of the explicit moltifier exhibited in the comment after conditions $(3.4)_4$.

Consider an initial condition $\omega \in C^1(\mathbb{R}^2)$ with compact support. For simplicity we assume also $\omega \geq 0$ and $\int \omega = 1$. In virtue of the estimates obtained at the end of Section 3, we have for the Euler flow $\omega \to \omega_t$:

$$\begin{cases} \sup_{t\in[0,T]} \text{diam}[\text{supp}\omega_t] \leq C \\ \sup_{t\in[0,T]} \|\nabla\omega_t\|_\infty \leq C \\ C^{-1}|\underline{x}_1-\underline{x}_2| \leq |\underline{x}_1(t)-\underline{x}_2(t)| \leq C|\underline{x}_1-\underline{x}_2| . \end{cases} \tag{5.18}$$

From now on we denote by C any constant depending only on ω and T.

Let us introduce a lattice in $\mathbb{R}^2$ spaced by $h > 0$ and denote by $\underline{x}_i$, $\underline{x}_j$ etc. its sites and let C_i be the cells of the lattice:

$$C_i = \{\underline{x} = (x^1,x^2) \in \mathbb{R}^2 |\ x_i^\alpha \leq x^\alpha \leq x_i^\alpha+h \quad \alpha=1,2\} \tag{5.19}$$

In each point of the lattice we put a vortex of intensity

$$\omega_i = \int_{C_i} \omega(\underline{y})d\underline{y} \tag{5.20}$$

i.e. we consider the vorticity distribution:

$$\omega^h(d\underline{x}) = \sum_j \omega_j \delta(\underline{x}_j) \tag{5.21}$$

We remark that the above summ is finite since ω is of compact support.

We want to compare the motion of the (regularized) vortex dynamics

$$\begin{cases} \dfrac{d\tilde{\underline{z}}_j(t)}{dt} = \underline{u}^h(\tilde{\underline{z}}_j(t),t) \equiv \sum\limits_{k\neq j} \omega_k \underline{k}_\varepsilon(\tilde{\underline{z}}_j(t) - \tilde{\underline{z}}_k(t)) \\ \tilde{\underline{z}}_j(0) = \underline{x}_j \end{cases} \tag{5.22}$$

with the "true" Euler evolution:

$$\begin{cases} \dfrac{d\underline{z}_j(t)}{dt} = \underline{u}(\underline{z}_j(t),t) \equiv \int d\underline{y}\, \underline{k}(\underline{z}_j(t)-\underline{y})\omega_t(\underline{y}) \\ \underline{z}_j(0) = \underline{x}_j\ . \end{cases} \tag{5.23}$$

Here $\varepsilon = \varepsilon(h)$ denotes, as above, a cutoff rate to be determined as function of h.

For a vector valued function $\{\underline{f}_j\}$ defined on the lattice sites, we define the following norm

$$\|\underline{f}_j\|_p = \Bigg(\sum_{\substack{j: \\ C_j \cap \operatorname{supp}\omega \neq \emptyset}} h^2 |\underline{f}_j|^p \Bigg)^{\frac{1}{p}} \tag{5.24}$$

The proof of the convergence is based on the following key proposition.

<u>Proposition 5.1</u>. Suppose for $h < \varepsilon < 1$ and $\bar{T} \le T$ we have:

$$\sup_j \sup_{t\in[0,\bar{T}]} |\underline{z}_j(t) - \tilde{\underline{z}}_j(t)| < \varepsilon\ . \tag{5.25}$$

Then, for $0 \le t \le \bar{T}$:

$$\|\underline{u}_j(t) - \underline{u}_t^h(t)\|_p \leq C(\|\underline{z}_j(t) - \tilde{\underline{z}}_j(t)\|_p - \varepsilon^2 \ln\varepsilon + \varepsilon^{-2}h) \tag{5.26}$$

where

$$\underline{u}_j(t) = \underline{u}(\underline{z}_j(t),t), \quad \underline{u}_j^h(t) = \underline{u}^h(\tilde{\underline{z}}_j(t),t) \tag{5.27}$$

<u>Proof</u>.

$$\underline{u}_j(s) - \underline{u}_j^h(s) = \underline{T}_1 + \underline{T}_2 + \underline{T}_3 \tag{5.28}$$

where

$$\int (\underline{k}(\underline{z}_j(s)-\underline{y}) - \underline{k}_\varepsilon(\underline{z}_j(s)-y))\omega_s(\underline{y})d\underline{y} = \underline{T}_1$$

$$\int (\underline{k}_\varepsilon(\underline{z}_j(s)-y) - \underline{k}_\varepsilon(\tilde{\underline{z}}_j(s)-y))\omega_s(y)d\underline{y} = \underline{T}_2 \tag{5.29}$$

$$\int \underline{k}_\varepsilon(\tilde{\underline{z}}_j(s)-y)[\omega_s(\underline{y})d\underline{y} - \omega_s^h(d\underline{y}) = \underline{T}_3$$

and where

$$\omega_t^h(d\underline{y}) = \sum_k \omega_k \delta_{\tilde{\underline{z}}_k(t)}(d\underline{y}) . \tag{5.30}$$

We estimate $\underline{T}_1$, $\underline{T}_2$, $\underline{T}_3$ separately.

$$|\underline{T}_1| = \left|\int_{|\underline{z}_j(s)-\underline{y}|<\varepsilon} d\underline{y}\{g(\underline{z}_j(s)-\underline{y}) - g_\varepsilon(\underline{z}_j(s)-\underline{y})\}(\underline{\nabla}^\perp\omega_s)(\underline{y})\right|$$

$$\leq -C\|\underline{\nabla}\omega_s\|_\infty \varepsilon^2 \ln\varepsilon \leq -C\varepsilon^2 \ln\varepsilon \tag{5.31}$$

by Green lemma, (5.18), $(3.4)_4$.

Hence

$$\|\underline{T}_1\|_p \leq -C\varepsilon^2 \ln\varepsilon \tag{5.32}$$

Still applying Green lemma we have

$$|\underline{T}_2| \leq \int_{\text{supp}\,\omega_t} dy\, |g_\varepsilon(\underline{z}_j(s)-\underline{y})-g_\varepsilon(\tilde{\underline{z}}_j(s)-\underline{y})|\, \|\underline{\nabla}\omega_s\|_\infty$$

$$\leq C|\underline{z}_j(s) - \tilde{z}_j(s)| \tag{5.33}$$

In fact by applying the mean value theorem we have to integrate $|(\underline{\nabla} g_\varepsilon)(\underline{z}'-\underline{y})|$ where $\underline{z}'$ is some point in the segment $\underline{z}_j(s)$, $\tilde{\underline{z}}_j(s)$. Such a function is bounded by $C\left(\frac{1}{|\underline{z}_j(s)-\underline{y}|}+\frac{1}{|\tilde{\underline{z}}_j(s)-\underline{y}|}\right)$ and hence (5.33). From this

$$\|\underline{T}_2\|_p \leq C\|\underline{z}_j(s) - \tilde{\underline{z}}_j(s)\|_p \tag{5.34}$$

It remains to estimate the last and more difficult term.

Defining

$$\tilde{\omega}_t^h(d\underline{y}) = \sum_j \omega_j \delta_{\underline{z}_j(t)}(d\underline{y}) \tag{5.35}$$

we have $\underline{T}_3 = \underline{S}_1 + \underline{S}_2$, where

$$\begin{aligned} \underline{S}_1 &= \int \underline{k}_\varepsilon(\tilde{\underline{z}}_j(s) - \underline{y})\{\omega_s(d\underline{y}) - \tilde{\omega}_s^h(d\underline{y})\} \\ \underline{S}_2 &= \int \underline{k}_\varepsilon(\tilde{\underline{z}}_j(s) - y)\{\tilde{\omega}_s^h(\underline{y})d\underline{y} - \omega_t^h(d\underline{y})\}. \end{aligned} \tag{5.36}$$

Let $\hat{P} \in C(\omega_s, \tilde{\omega}_s^h)$, then

$$|\underline{S}_1| = \inf_{\hat{P}} \left|\int (\underline{k}_\varepsilon(\tilde{\underline{z}}_j(s) - y) - \underline{k}_\varepsilon(\tilde{\underline{z}}_j(s) - \underline{z}))\hat{P}(d\underline{y}, d\underline{z})\right|$$

$$\leq C\varepsilon^{-2} R(\omega_s, \tilde{\omega}_s^h) \tag{5.37}$$

Given $\hat{Q} \in C(\omega,\omega^h)$, we define $\hat{Q}_t \in C(\omega_t,\tilde{\omega}_t^h)$ via

$$\int \hat{Q}_t(d\underline{x},d\underline{y})f(\underline{x},\underline{y}) = \int \hat{Q}(d\underline{x},d\underline{y})f(\underline{x}(t),\underline{y}(t)) \tag{5.38}$$

where $\underline{x}(t)$ and $\underline{y}(t)$ are characteristics of the Euler flow starting from $\underline{x}$ and $\underline{y}$ at time zero. By (5.18)

$$\int \hat{Q}_t(d\underline{x},d\underline{y})|\underline{x}-\underline{y}| \leq C\int \hat{Q}(d\underline{x},d\underline{y})|\underline{x}-\underline{y}| \tag{5.39}$$

and hence:

$$R(\omega_s,\tilde{\omega}_s^h) \leq CR(\omega,\omega^h). \tag{5.40}$$

Defining

$$\hat{Q}(d\underline{x},d\underline{y}) = \sum_j \delta_{x_j}(d\underline{x})\omega(\underline{y})\chi(\underline{y} \in C_j)d\underline{y} \tag{5.41}$$

we verify easily that $\hat{Q} \in C(\omega,\omega^h)$ and hence

$$R(\omega,\omega^h) = \int \hat{Q}(d\underline{x},d\underline{y})|\underline{x}-\underline{y}| \leq Ch \tag{5.42}$$

In conclusion

$$\|\underline{S}_1\|_p \leq C\varepsilon^{-2}h \tag{5.43}$$

Finally we estimate $\underline{S}_2 = (S_2^1,S_2^2)$.

$$S_2^\alpha = \sum_k{}' (k_\varepsilon^\alpha(\tilde{\underline{z}}_j(s) - \tilde{\underline{z}}_k(s)) - k_\varepsilon^\alpha(\tilde{\underline{z}}_j(s) - z_k(s)))\omega_k$$

$$= \sum_k{}' (\underline{\nabla} k_\varepsilon^\alpha)(\underline{z}_j(s)-\underline{z}_k(s)+\underline{y}_{jk}^\alpha)\cdot(\tilde{\underline{z}}_k(s)-\underline{z}_k(s))\omega_k \tag{5.44}$$

where $\sum_k{}'$ means $\sum_{\substack{k: \\ \mathrm{supp}\,\omega \cap C_k \neq \emptyset}}$, $\alpha = 1,2$, and $\underline{y}^{\alpha}_{jk}$ is a point in the segment $(\tilde{\underline{z}}_j - \underline{z}_j, (\tilde{\underline{z}}_j - \underline{z}_j) + (\underline{z}_k - \tilde{\underline{z}}_k))$. By hypothesis $|\underline{y}^{\alpha}_{jk}| < \varepsilon$.

We write the above expression in the following way

$$S^{\alpha}_2 = \int \underline{K}(\underline{z},\underline{z}')\cdot\underline{f}(\underline{z}')d\underline{z}' \tag{5.45}$$

where

$$\underline{K}(\underline{z},\underline{z}') = \underline{K}_1(\underline{z},\underline{z}') + \underline{K}_2(\underline{z},\underline{z}')$$

$$\begin{aligned} \underline{K}_1(\underline{z},\underline{z}') &= \underline{\nabla} k^{\alpha}_{\varepsilon}(\underline{z}_j(s) - \underline{z}_k(s) + \underline{y}^{\alpha}_{jk}) - \underline{\nabla} k^{\alpha}_{\varepsilon}(\underline{z}-\underline{z}') \\ &\quad \text{if } \underline{z} \in C_j(s), \quad z' \in C_k(s) \end{aligned} \tag{5.46}$$

$$\underline{K}_2(\underline{z},\underline{z}') = \underline{\nabla} k^{\alpha}_{\varepsilon}(\underline{z}-\underline{z}')$$

where $C_j(s)$ is the cell C_j evolved under the Euler flow and where:

$$\underline{f}(z) = (\tilde{\underline{z}}_k(s) - \underline{z}_k(s))\omega_k h^{-2} \quad \text{if} \quad z \in C_k \tag{5.47}$$

To estimate $\underline{K}_1$ we observe that, denoting by D^2 any 2nd derivative of k^{α}_{ε}, one has

$$\sum_j{}' \sup_{|\underline{y}_{jk}| \leq C_o\varepsilon} (D^2 k^{\alpha}_{\varepsilon})(\underline{z}_j(s) - \underline{z}_k(s) + \underline{y}_{jk})h^2 \leq C\varepsilon^{-1}. \tag{5.48}$$

The above estimate follows from the regularity of the Euler motion. For details see [Ha.2] Lemma 5 or [Bea.2] Lemma 3.2. Applying the mean value theorem in the definition of $\underline{K}_1$ and estimate (5.48) we obtain (using the bound $\mathrm{diam}\, C_j(t) \leq \mathrm{diam}\, C_j \leq Ch \leq C\varepsilon$)

$$\max\left(\int |\underline{K}_1(\underline{z},\underline{z}')|\,d\underline{z}',\ \int |\underline{K}_1(\underline{z},\underline{z}')|\,d\underline{z}\right) \leq C\varepsilon\varepsilon^{-1} = C \tag{5.49}$$

and hence by Hölder inequality (p and q are conjugate exponents)

$$|\underline{K}_1 f|(\underline{z}) \leq C^{\frac{1}{q}}\left(\int |\underline{K}_1(\underline{z},\underline{z}')|\,|\underline{f}(\underline{z}')|^p d\underline{z}'\right)^{\frac{1}{p}} \tag{5.50}$$

and finally

$$\|\underline{K}_1 f\|_p \leq C^{1/q}\, C^{1/p} \|f\|_p \leq C\|\underline{f}\|_p \tag{5.51}$$

yielding:

$$\|\underline{K}_1 f\|_p \leq C\|\tilde{\underline{z}}_k(s) - \underline{z}_k(s)\|_p \tag{5.52}$$

To estimate K_2 we observe first that

$$\|\underline{\nabla k}^{\alpha}_{\varepsilon} * \underline{f}\|_p \leq C\|\underline{f}\|_p \tag{5.53}$$

Inequality (5.53) follows by the inequality $\|\underline{\nabla k}^{\alpha} * \underline{f}\|_p \leq C\|\underline{f}\|_p$ (see [St.1] p. 39), and by Young inequality $\|\rho_\varepsilon * \underline{f}\|_p \leq C\|\underline{f}\|_p$.

Thus:

$$\|\underline{K}_2 \underline{f}\|_p \leq C\|\tilde{\underline{z}}_k(s) - \underline{z}_k(s)\|_p\ . \tag{5.54}$$

This conclues the proof of the Proposition. □

<u>Theorem 5.2</u>. We choose $\varepsilon = h^{1/4}$ then:

$$\|\underline{z}_j(t) - \tilde{\underline{z}}_j(t)\|_p \leq -C\sqrt{h}\ln h \tag{5.55}$$

for a sufficiently large p and small h < 1.

Proof. Let $\bar{T} < T$ be a time for which $\sup_{t \leq \bar{T}} |\underline{z}_j(t) - \tilde{\underline{z}}_j(t)| < \varepsilon$. Then

$$\frac{d}{dt}\|\underline{z}_j(t) - \tilde{\underline{z}}_j(t)\|_p \leq \|\underline{u}_j(t) - u_j^h(t)\|_p \leq C\|\underline{z}_j(t) - \tilde{\underline{z}}_j(t)\|_p - C\sqrt{h}\,\ln h \tag{5.56}$$

and hence up to time $\bar{T}$:

$$\|\underline{z}_j(t) - \tilde{\underline{z}}_j(t)\|_p \leq -C\sqrt{h}\,\ln h \tag{5.57}$$

But if T^* is the first time s.t. $|\tilde{z}_j(T^*) - z_j(T^*)| = \varepsilon$ for some j, then:

$$\|\underline{z}_j(T^*) - z_j(T^*)\|_p \geq h^{\frac{2}{p}}|z_j(T^*) - z_j(T^*)| = h^{\frac{2}{p}+\frac{1}{4}} \tag{5.58}$$

that makes impossible the inequlity (5.57) when $T^* < T$, $p > 8$, h suitably small. Hence the hypothesis of Proposition 5.1 are verified and proof of Theorem 5.2 achieved. □

The results of Theorem 4.1 can be considerably improved along the lines of Proposition 1 but with a larger technical effort in estimating T_1 and T_2.

In this point it is crucial the choice of the cutoff. With the cutoff choosen in a suitable class one can obtain higher power in h in the estimate of the error $\|z_j - \tilde{z}_j\|_p$. See [Ha.2] and [Bea.1].

Notes

The convergence of the vortex model to the Euler evolution is a relatively widely investigated problem due to the interest of finding converging algorithms. See [Sa.1].

A previous result, valid for short times, can be found in [Ha.1]. Results valid for all times and with careful estimates of the error, are in [Ha.2], [Bea.1]. Theorem 4.1 is taken by [Ma.1]. It makes use of techniques employed in [Do.1] in which, similar problems where treated in the context of the Vlasov equation. Vlasov-like dynamics has been investigated also by other authors [Mc.2], [Br.1], [Ne.1]. In all these works it was assumed smoothness of the kernel, making the results not directly applicable to the Euler case.

For a review concerning practical aspects of numerical simulation of flows by means of vortex methods see [Le.1].

6. NAVIER-STOKES EQUATION

Although the Euler equations describe quite well the behavior of a fluid in many circumstances, there are situations in which is inadeguate to explain features of real fluids. For instance it is well known the D'Alembert's paradox for which the force acting on a body moving in an irrotational stationary ideal fluid, is zero.

There are many discrepances between the ideal evolution and the experience mostly due to the too schematic description of the interactions between the fluid and the wall. In particular there is not any mechanism in the Euler equation for the production of vorticity for an ideal fluid in a container and such an evolution does not provide a diffusion of the vorticity as really happens. These inconsistences can be overcome by a more careful treatment of the stress tensor. In Section 1 we supposed that the only non zero contribution arises from the diagonal part. This assumption means that we neglected the friction between different layers of fluid. We now modify the expression of the stress tensor by taking into account also this fact. Let us introduce the deformation tensor

$$D_{ij} = \frac{1}{2}\left(\frac{\partial u_i}{\partial x_j} + \frac{\partial u_j}{\partial x_i}\right) \qquad (6.1)$$

that gives a measure of the deformation of the fluid. Assuming the first correction to the stress tensor to be proportional (linear approximation) to the deformation, we have ($\rho = 1$)

$$\begin{cases} \dfrac{D\underline{u}}{Dt} = -\underline{\nabla p} + \nu\Delta u \\ \underline{\nabla \cdot u} = 0 \end{cases} \qquad \text{(Navier-Stokes equations)} \qquad (6.2)$$

where $\nu > 0$ is called viscosity coefficient.

Togheter with Eq.s (6.2) we have to fix the boundary conditions that describe the interaction between fluid and walls. It is usually assumed

$$\underline{u} = 0 \quad \text{on} \quad \partial D, \tag{6.3}$$

that means perfect adherence of the fluid on the boundary.

If D is unbounded as before we require $\underline{u}(\underline{x}) \to 0$ when $|\underline{x}| \to +\infty$, $\underline{x} \in D$.

The Navier-Stokes equation is more realistic than the Euler equation in all cases in which the friction is not negligible and explains the phenomena mentioned above. In particular it can describe the production of vorticity near the wall. We shall discuss in more detail this point in Section 9.

The Euler and Navier-Stokes equations are different for two reasons. Firstly they differs by the term $\nu\Delta\underline{u}$. Secondly they obey to different boundary conditions.

Let us see if it is reasonable to expect that Navier-Stokes and Euler solutions are "near" when ν is small. Let us first clarify what means ν small. Let L and U be characteristic lenght and velocity of the fluid. Defining the following dimensionaless new variables

$$\underline{u}' = \underline{u}/U, \quad \underline{x}' = \underline{x}/L, \quad t' = t/\tau \quad \text{where} \quad \tau = L/U, \tag{6.4}$$

Navier-Stokes equation becomes

$$\frac{D\underline{u}'}{Dt'} = -\underline{\nabla}'p' + \frac{1}{R}\Delta'\underline{u}' \tag{6.5}$$

where

$$p' = p/U^2, \qquad \underline{\nabla}' = \left(\frac{\partial}{\partial x_i'}\right)_{i=1}^{n}, \qquad R = \frac{LU}{\nu}. \tag{6.6}$$

R is called Reynolds number.

It is clear that two fluids with the same Reynolds number moving in the same domain are similar.

From the above discussion the Euler solutions are expected to be approached by Navier-Stokes solutions for large values of R. Such limit is delicate because of the singular character of the perturbation $\frac{1}{R}\Delta\underline{u}$ (containing highest order derivatives) and mostly for the difference of the boundary conditions. For the last reason it is certainly true that the Navier-Stokes solution cannot converge to the Euler solution in a region at distance δ from the boundary, where δ is a suitable function of R. Such region is called boundary layer. Let us give an idea of the thickness of the boundary layer by considering an explicit example.

$$D = \{x_1, x_2 | x_2 \geq 0\} \qquad \underline{u}(x_1, +\infty, t) = (U,0) \tag{6.7}$$

Let us look for a solution such that

$$\begin{aligned} &\underline{u}(x_1,x_2,t) = (u_1(x_2,t),0) \\ &p = \text{const.} \end{aligned} \tag{6.8}$$

The Euler equation has as unique solution $\underline{u} = (U,0)$. The Navier-Stokes equation is solved by:

$$\begin{cases} u_1(x_1,x_2,t) = \dfrac{2U}{\sqrt{\pi}} \displaystyle\int_0^{\frac{x_2}{2\sqrt{\nu t}}} (\exp - s^2)ds \\ u_2(x_1,x_2,t) = 0 \end{cases} \tag{6.9}$$

The two solutions are sensibly different in a strip of thickness $\sqrt{\nu t}$ and hence

$$\delta = \sqrt{\nu\tau} = \frac{\tau U}{\sqrt{R}} \tag{6.10}$$

where τ is some characteristic time.

The vorticity associated to the solution (6.9) is

$$\omega(x_1,x_2;t) = -\frac{U}{\sqrt{\pi\nu t}} e^{-\frac{x_2^2}{4\nu t}} . \tag{6.11}$$

The physical situation described in the above example is the following. At time zero we have the initial datum $\underline{u} = (U,0)$ that remains constant in time for the Euler evolution. The friction between the fluid and the wall in the Navier-Stokes case creates a singular distribution of vorticity on the boundary at time zero. At later times it diffuses following the heat equation since in this case the non linear terms vanish.

We mention two approximations to the Navier-Stokes equations. The first one consists in neglecting the non linear term w.r.t. the diffusive term:

$$\frac{\partial \underline{u}}{\partial t} = -\underline{\nabla} p + \nu\Delta\underline{u} \quad \text{(Stokes equation)} \tag{6.12}$$

Obviously eq. (6.12) is expected to be realistic for low Reynolds number and in situations in which gradients of velocity are small.

The second one consists in studying Navier-Stokes equations in the boundary layer region. Taking into account only zero order term in δ (after the scaling $y \to y\delta$, $u_2 \to u_2\delta$, $R \to R/\delta^2$) we have

$$\begin{cases} \dfrac{\partial u_1}{\partial t} + u_1\dfrac{\partial u_1}{\partial x_1} + u_2\dfrac{\partial u_1}{\partial x_2} = -\dfrac{\partial P}{\partial x_2} + \dfrac{1}{R}\dfrac{\partial^2 u_1}{\partial x_2^2} \\ \dfrac{\partial P}{\partial x_2^2} = 0 \\ \dfrac{\partial u_1}{\partial x_1} + \dfrac{\partial u_2}{\partial x_2} = 0 \qquad x_1 \in \mathbb{R},\quad x_2 \geq 0. \end{cases} \tag{6.13}$$

The solutions of (6.13) have to be matched with the solutions of the "main flow" (the flow at infinity) satisfying the Euler equation

$$\frac{\partial V}{\partial t} + V\frac{\partial V}{\partial x_1} = -\frac{\partial P}{\partial x_1} \tag{6.14}$$

Thus the boundary conditions associated to the initial value problem (5.13) are:

$$\begin{aligned} &u_1(x_1,0) = 0 \qquad u_2(x_1,0) = 0 \\ &\lim_{x_2\to\infty} u_1(x_1,x_2) = V(x_1) \end{aligned} \tag{6.15}$$

Eq.s (6.13) are called Prandtl equations and are expected to give a good description in case of slightly viscous fluid and in regions very near to the wall.

From a mathematical point of view Prandtl equations are

rather difficult to deal with. Only local existence theorems are available till now. Presumibly in finite time the solution may develop singularities in contrast with the hypothesis under which the equation itself was derived.

We conclude this section by discussing the Navier-Stokes equations in terms of vorticity in analogy with previous analysis. In two dimensions we have:

$$\left(\frac{\partial}{\partial t} + \underline{u}\cdot\nabla\right)\omega = \nu\Delta\omega \tag{6.16}$$

When $D = \mathbb{R}^2$ or T^2 a theory in analogy with the Euler case can be developped. The only difference is that when the motion of an element of vorticity is considered, a diffusion has to be added taking into account the Laplacian in Eq. (6.16). Ordinary differential equations have to be replaced by stochastic ones, as we shall see in the next Section, but the main ideas are the same. When boundaries are present the situation is much more involved: there are not simple boundary conditions to impose on the vorticity in such a way that $\underline{u}$, constructed via the Dirichlet Green function, satisfies the right boundary conditions. This problem will be discussed in Section 9.

Notes

Basic facts about viscous fluids can be found in the literature quoted in Section 1. In particular in [VM.1] it is extensively discussed the Prandtl limit in the stationary case. Useful change of variables are introduced reducing the stationary Prandtl problem to the solution of an ordinary differential equation.

Chorin proposed a numerical algorithm for the time dependent case [Ch.5].

O. Oilenik obtained local existence and uniqueness results for time dependent Prandtl equation [O.2].

In [Be.1] the same result is obtained by means of a stochastic approach, similar to that discussed in the next Section.

For existence and uniqueness of the stationary solution see [O.1].

A rigorous deduction of the stationary Prandtl equation has been obtained in [Fi.1]. See also [Se.1].

7. DIFFUSION PROCESS AND NAVIER-STOKES EQUATIONS

In this Section we want to establish a connection between the Navier-Stokes equations in $\mathbb{R}^2$ and a certain class of diffusion processes. Such connection will provide a good understanting of the mathematical structure of the Navier-Stokes equation in two dimensions and suggests practical methods to approximate its solutions.

We begin by giving some elementary facts about diffusion processes.

Let us consider a stochastic process in $\mathbb{R}^\nu$ i.e. a family $\{\underline{x}(t)\}_{t\in\mathbb{R}}$ of $\mathbb{R}^\nu$ valued random variables defined in some probability space $(\Omega,\Sigma,\mathbb{P})$. As usual Ω is a set, Σ a σ-algebra of subset of Ω, and $\mathbb{P}$ a probability measure defined on it. We shall suppose that such process is described by a family of transition probabilities $p(x,s;dy,t)$ i.e.

$$p(s,\underline{x};A,t) = \mathbb{P}\{\underline{x}(t) \in A|\underline{x}(s) = \underline{x}\} \tag{7.1}$$

for all borel set $A \subset \mathbb{R}^\nu$, $\underline{x} \in \mathbb{R}^\nu$, $t > s$. The r.h.s. of (7.1) denotes the probability that the process belongs to A at time t, knowing it is at $\underline{x}$ at time s.

Example 1 (ν-dimensional brownian motion). A ν-dimensional brownian motion $\underline{b}(t) = \{b_i(t)\}_{i=1,\nu}$ is a family of $\mathbb{R}^\nu$-valued gaussian random variables with mean zero and covariance given by:

$$\mathbf{E}(b_i(t)b_j(s)) = \delta_{ij}\min(t,s) \tag{7.2}$$

Some elementary consequences of this definition are

a) $\mathbf{E}(|\underline{b}(t) - \underline{b}(s)|^2) = |t-s|$

b) $\underline{b}(t) - \underline{b}(s)$ is independent of $\underline{b}(s)$, $s < t$ and is still a brownian motion.

c) $$p(s,\underline{x};d\underline{y},t) = \frac{d\underline{y}}{[2\pi(t-s)]^{\frac{\nu}{2}}} \exp\left[-\frac{(\underline{x}-\underline{y})^2}{2(t-s)}\right] \tag{7.3}$$

Moreover $\underline{b}(t)$ has continuous paths in the sense that the probability that $\underline{b}(t)$ is continuous is one.

The brownian motion is connected with the heat equation as follows. Let μ_s be a Borel probability measure in $\mathbb{R}^\nu$ describing the distribution of the "brownian particle" at time $s < t$. Then the distribution at time $t > s$ is given by the formula:

$$\mu_t(d\underline{y}) = \int \mu_s(d\underline{x}) p(\underline{x},s;d\underline{y}t) \tag{7.4}$$

By an explicit calculation:

$$\frac{d}{dt} \mu_t(f) = \frac{1}{2} \mu_t(\Delta f) \tag{7.5}$$

for all bounded C^2 functions f.

Example 2 (Deterministic motion). Let us consider the differential equation

$$\dot{\underline{x}} = \underline{a}(\underline{x},t) \tag{7.6}$$

where $\underline{a}(\cdot,t)$ is a sufficiently smooth, time dependent vector field.

If $\underline{x}(t;\underline{x},s)$ is the solution of the initial value problem (7.6) with initial condition $\underline{x}(s;\underline{x},s) = \underline{x}$, then such solution may be interpreted as a process whose transition probabilities are given by:

$$p(\underline{x},s;d\underline{y},t) = \delta(\underline{y} - \underline{x}(t;\underline{x},s))d\underline{y} \ . \tag{7.7}$$

If μ_s is the distribution of $\underline{x}$ at time s then

$$\mu_t(f) = \int \mu_s(d\underline{x}) f(\underline{x}(t;\underline{x},s)) \tag{7.8}$$

$$\begin{aligned}\frac{d\mu_t}{dt}(f) &= \int \mu_s(d\underline{x})(\underline{\nabla} f)(\underline{x}(t;\underline{x},s)) \cdot \dot{\underline{x}}(t;\underline{x},s) \\ &= \int \mu_s(d\underline{x}) \underline{a}(\underline{y},t) \cdot (\underline{\nabla} f)(\underline{y}) \delta(\underline{y} - \underline{x}(t;\underline{x},s)) d\underline{y} \\ &= \mu_t(\underline{a} \cdot \underline{\nabla} f) \ . \end{aligned} \tag{7.9}$$

The above two examples suggest to interpret the following class of parabolic linear equations, ($\sigma \in \mathbb{R}$):

$$\frac{\partial}{\partial t}\mu(\underline{x},t) = -[\underline{\nabla} \cdot \underline{a}\mu](\underline{x},t) + \frac{\sigma^2}{2} \Delta\mu(\underline{x},t) \qquad (7.10)_a$$

or in a weak form

$$\frac{d}{dt}\mu_t(f) = \mu_t(\underline{a} \cdot \underline{\nabla} f) + \frac{\sigma^2}{2}\mu_t(\Delta f), \qquad (7.10)_b$$

as an evolution equation for the distribution density μ of a process that is a combination of a brownian motion and a deterministic process generated by the time dependent vector field $\underline{a}$.

Such a process may be constructed in the following way. Consider a brownian motion $\underline{b}(t)$ realized on some probability space $(\Omega,\Sigma,\mathbb{P})$. For each (continuous) sample $\underline{b}(t)[\xi]$, $\xi \in \Omega$ of the brownian motion, let $x(t)[\xi]$ be the solution of the integral equation:

$$\underline{x}(t)[\xi] = \underline{x} + \sigma\underline{b}(t)[\xi] + \int_s^t dr\, \underline{a}(\underline{x}(r)[\xi],r) \quad . \tag{7.11}$$

Such solution exists unique under suitable assumption for $\underline{a}$, since $\underline{b}(t)$ is continuous. It is easily seen that the process defined via the mapping (almost everywhere defined)

$$\Omega \ni \xi \to \underline{x}(t)[\xi] \tag{7.12}$$

has the following properties when $\underline{a}$ is bounded with its first derivaties:

a) $\underline{x}(t)$ has continuous paths with probability one.

b) $\mathbf{E}(\underline{x}(t+\Delta t)-\underline{x}(t)|\underline{x}(t)) = \underline{a}(\underline{x}(t),t)\Delta t+o(\Delta t)$

(7.13)

c) $\mathbf{E}(x_i(t+\Delta t)-x_i(t))(x_j(t+\Delta t)-x_j(t))|\underline{x}(t)) =$
$= \delta_{ij}\sigma^2\Delta t + o(\Delta t)$

Here $\mathbf{E}(|)$ denotes as usual, the conditional expectation.

The process we have defined is a particular case of a more general class of processes, called diffusion processes.
The general definition allows space and time dependence of the diffusion constant σ. This would complicate our analysis, requiring the concept of the stochastic integration. Since this machinery is not necessary to our fluid dynamical purposes, we avoid it here.

Let us come back to the evolution equation (7.10). Consider a process satisfying (7.11) or in a short hand notation

$$\begin{cases} d\underline{x}(t) = \underline{a}(\underline{x}(t),t)dt + \sigma d\underline{b}(t) \\ \underline{x}(s) = \underline{x} \quad \text{almost surely}, \quad s < t. \end{cases} \tag{7.14}$$

(7.14) is called a stochastic differential equation with drift $\underline{a}$ and diffusion constant σ.

Defining the transition probabilities of the process via the formula:

$$\int p(\underline{x},s;d\underline{y},t)f(\underline{y}) = \mathbf{E}[f(\underline{x}(t;x,s))] \tag{7.15}$$

where $\underline{x}(t;\underline{x},s)$ satisfies (7.14), the evolved distributions are defined as:

$$\mu_t(d\underline{y}) = \int \mu_s(d\underline{x})p(\underline{x},s;d\underline{y},t) \tag{7.16}$$

Then, if f is sufficiently smooth:

$$\begin{aligned} &\mathbf{E}(f(\underline{x}(t+\Delta t;\underline{x},s) - f(\underline{x}(t;\underline{x},s)) = \\ &\quad \mathbf{E}((\underline{\nabla} f)(\underline{x}(t;\underline{x},s))\cdot(\Delta\underline{x})(t)) + \tfrac{1}{2}\mathbf{E}(\Sigma\frac{\partial^2 f}{\partial x_i \partial x_j}\Delta_i x \Delta_j x(t)) + o(\Delta t) \end{aligned} \tag{7.17}$$

where $\Delta\underline{x} \equiv \{x(t+\Delta t;\underline{x},s)_i - x(t;\underline{x},s)_i\}_{i=1,\nu} = \{\Delta_i x(t)\}_{i=1,\nu}$.

In virtue of (7.13):

$$\begin{aligned} &\mathbf{E}((\underline{\nabla} f)(\underline{x}(t;\underline{x},s))\cdot(\Delta\underline{x})(t)) = \\ &= \mathbf{E}((\underline{\nabla} f)(\underline{x}(t;\underline{x},s)\cdot\mathbf{E}((\Delta\underline{x})(t)|\underline{x}(t))) \doteq \\ &= \mathbf{E}((\underline{\nabla} f)(\underline{x}(t;\underline{x},s)))\cdot\underline{a}(\underline{x}(t),t)\Delta t + o(\Delta t) \end{aligned}$$

$$\mathbf{E}(\Sigma\frac{\partial^2 f}{\partial x_i \partial x_j}\,\Delta_i x \Delta_j x) = \mathbf{E}(\Sigma\frac{\partial^2 f}{\partial x_i \partial x_j}(\underline{x}(t;\underline{x},s))\mathbf{E}(\Delta_i x \Delta_j x|\underline{x}(t))) =$$

$$= \mathbf{E}[(\Delta f)(\underline{x}(t;\underline{x},s))]\sigma^2 \Delta t + o(\Delta t) \qquad (7.18)$$

and hence, by a straightforward calculation, we derive $(7.10)_b$

$$\frac{d}{dt}\mu_t(f) = \mu_t[(\underline{a}\cdot\underline{\nabla})f] + \frac{\sigma^2}{2}\mu_t(\Delta f) \; . \qquad (7.19)$$

If μ_t has a density $\mu_t(\underline{x})$ and such a density is $C^2(\mathbb{R}^\nu)$, then (7.19) implies $(7.10)_a$.

The above considerations may be specialized in the context of Navier-Stokes equations written in terms of vorticity:

$$\begin{cases} \frac{\partial \omega}{\partial t}(\underline{x},t) + (\underline{u}\cdot\underline{\nabla})\;(\underline{x},t) = \nu\Delta\omega \\ \omega = \text{curl}\;\underline{u}, \quad \underline{\nabla}\cdot\underline{u} = 0 \; . \end{cases} \qquad (7.20)$$

Suppose $\underline{u} = \underline{u}(\underline{x},t)$ be a sufficiently smooth solution of the Navier-Stokes equations. Consider the process $\underline{x}(t;\underline{x})$ solution of the following stochastic differential equation

$$\begin{cases} d\underline{x}(t) = \underline{u}(\underline{x}(t),t)dt + \sigma d\underline{b}(t) \\ \underline{x}(0) = \underline{x} \text{ almost surely, } \frac{\sigma^2}{2} = \nu \end{cases} \qquad (7.21)$$

then, because $\underline{\nabla}\cdot\underline{u} = 0$,

$$\int \omega(\underline{x},t)f(\underline{x})d\underline{x} = \int \omega(\underline{x},0)\mathbf{E}(f(\underline{x}(t;\underline{x}))d\underline{x} \qquad (7.22)$$

where $\omega(\underline{x},t) = \text{curl}\,\underline{u}(\underline{x},t)$, $t \geq 0$ satisfies (7.20).

Now we are able to construct the solutions of the Navier-Stokes equation in $\mathbb{R}^2$ following step by step all arguments lead ing to the construction of the Euler flow, just replacing the ordinary differential problem $\dot{\underline{x}} = \underline{u}(\underline{x},t)$, by the stochastic differential problem $d\underline{x}(t) = u(\underline{x}(t),t)dt + \sigma d\underline{b}(t)$, and by taking expectations when necessary.

Physically speaking the effect of the viscosity may be thought as a stochastic perturbation: each element of vorticity moves accordingly to the velocity field $\underline{u}$, but in addition a random motion is present, that delocalizes it and hence influences the velocity field at later times.

The above ideas can be summarized in the following theorem.

Theorem 7.1. Let $\omega \in M(a,b)$. There exists a unique signed measure valued function $t \to \omega_t^\varepsilon$ satisfying the initial value problem

$$\begin{cases} \frac{d}{dt}\omega_t^\varepsilon(f) = \omega_t^\varepsilon(\underline{u}^\varepsilon \cdot \underline{\nabla} f) + \nu\omega_t^\varepsilon(\Delta f) \\ \omega_o^\varepsilon = \omega \end{cases} \tag{7.23}$$

$$\underline{u}^\varepsilon(\underline{x},t) = \int \underline{k}_\varepsilon(\underline{x}-\underline{y})\omega_t^\varepsilon(\underline{y})d\underline{y}, \quad \varepsilon > 0 \tag{7.24}$$

If in addition $\omega(d\underline{x}) = \omega(\underline{x})d\underline{x}$ with $\omega \in L_1 \cap L_\infty(\mathbb{R}^2)$

$$\lim_{\varepsilon \to 0} R(\omega_t^\varepsilon, \omega_t) = 0$$

and $\omega_t(dx) = \omega_t(x)dx$, $\omega_t \in L_1 \cap L_\infty(\mathbb{R}^2)$ is the unique solution of the initial value problem (7.23), (7.24) with $\varepsilon = 0$.

Finally a stochastic process $\underline{x}^\varepsilon(t;x)$ may be constructed as solution of $(\varepsilon \geq 0)$

$$dx^\varepsilon(t) = \underline{u}^\varepsilon(x^\varepsilon(t),t)dt + \sigma d\underline{b} \qquad \frac{\sigma^2}{2} = \nu \tag{7.25}$$

such that:

$$\omega_t^{\varepsilon}(f) = \int \omega(d\underline{x})\mathbb{E}f(\underline{x}^{\varepsilon}(t;\underline{x})) \tag{7.26}$$

Proof (Sketch). One can follow the lines of the proofs of Thms. (3.1) and (3.2).

First one defines a stochastic process solution of

$$d\underline{x}^{\varepsilon,\mu}(t) = \underline{u}^{\varepsilon,\mu}(\underline{x}^{\varepsilon,\mu}(t),t)dt + \sigma d\underline{b} \tag{7.27}$$

where

$$\underline{u}^{\varepsilon,\mu}(\underline{x},t) = \int \underline{k}_{\varepsilon}(\underline{x}-\underline{y})\mu_t(dy) . \tag{7.28}$$

Then the map

$$\mu_t(f) \to \int \omega(d\underline{x})\mathbb{E}[f(\underline{x}^{\varepsilon,\mu}(t;\underline{x}))] \tag{7.29}$$

is well defined and has a unique fixed point for $\varepsilon > 0$.

The cutoff can be removed as in Section 3 Th. (3.2). □

Remark 1. We notice that a datum, initially in $M(a,b)$, under the action of the Euler evolution, remains in such space at later times because of the invariance of the measure and uniqueness of the characteristics. This is no more true for the Navier-Stokes evolution. Stochastic evolutions allows a superposition of positive and negative part of the measure as in the case of the heat equation.

Remark 2. A datum initially in $L_1 \cap L_\infty$ increases its regularity in time. Actually it is possible to show that ω_t solving the initial value problem (7.23) for $\varepsilon = 0$, it C^∞ for positive times.

What follows by our analysis is that $\omega_t \in L_1 \cap L_\infty$ and this means that u is bounded and Hölder continuous.

Applying perturbative techniques used in the theory of linear parabolic equation [Fri.1] one can show, via a bootstrap argument, the assertion.

The stochastic formalism introduced above, allows us to prove the convergence of a Navier-Stokes solution to the corresponding Euler solution in the limit $\nu \to 0$. Obviously, these techniques does not work in presence of boundary. As far as we know this interesting problem is still open.

Theorem 7.2. Let ω_t^ν, $t \geq 0$, be the solution of the Navier-Stokes equation (7.20), with initial condition $\omega_o \in L_1 \cap L_\infty$.

Then, for all continuous bounded f:

$$\lim_{\nu \to 0} \omega_t^\nu(f) = \omega_t(f) \tag{7.30}$$

where ω_t is the weak solution of the Euler equation with initial datum ω_o.

Proof. We assume $\omega_o \geq 0$. Let $\underline{x}^\nu(t)$ be the solution of

$$\underline{x}^\nu(t) = \int_0^t \underline{u}^\nu(\underline{x}^\nu(s),s)ds + \sigma\underline{b}(t) + \underline{x}, \qquad \sigma = \sqrt{2\nu} \tag{7.31}$$

where

$$\underline{u}^\nu(\cdot,t) = -\underline{\nabla}^\perp \Delta^{-1} \omega_t^\nu \tag{7.32}$$

and $\underline{x}(t)$ the solution of

$$\underline{x}(t) = \int_0^t \underline{u}(\underline{x}(s),s) + \underline{x} \tag{7.33}$$

where

$$\underline{u}(\cdot,t) = -\underline{\nabla}^{\perp}\Delta^{-1}\omega_t \,. \tag{7.34}$$

We have

$$\begin{aligned}|\underline{x}^{\nu}(t)-\underline{x}(t)| \le & \int_0^t ds\int|k(\underline{x}^{\nu}(s)-\underline{y})-\underline{k}(\underline{x}(s)-\underline{y})|\omega_s(\underline{y})d\underline{y} \\ & +\int_0^t ds\left|\int\underline{k}(\underline{x}^{\nu}(s)-\underline{y})\{\omega_s^{\nu}(\underline{y})-\omega_s(\underline{y})\}d\underline{y}\right| + \sqrt{2\nu}\,\underline{b}(t)\end{aligned} \tag{7.35}$$

Then, denoting by $\mathbf{E}_x$ the conditional expectation to the event $\underline{x}^{\nu}(0) = \underline{x}(0) = \underline{x}$,

$$\begin{aligned}\int\omega(\underline{x})\mathbf{E}_x(|\underline{x}^{\nu}(t)-\underline{x}(t)|)d\underline{x} \le & \, C\int_0^t ds\int\omega(\underline{x})\mathbf{E}_x(\varphi(\underline{x}^{\nu}(s),\underline{x}(s)))d\underline{x} + \\ & \int_0^t ds\int d\underline{x}\omega(\underline{x})\mathbf{E}_x\left|\int\underline{k}(\underline{x}^{\nu}(s)-\underline{y})\{\omega_s^{\nu}(\underline{y}) -\omega_s(\underline{y})|\}d\underline{y}\right| + \sqrt{2\nu}\|\omega_0\|_1\sqrt{t}\end{aligned} \tag{7.36}$$

where φ is defined in (3.29) and we have used the inequality $(3.28)_2$ and $\|\omega_t^{\nu}\|_1 \le \|\omega_0\|_1$ and $\|\omega_t^{\nu}\|_\infty \le \|\omega_0\|_\infty$.

The second integral in the r.h.s. of (7.36) becomes

$$\begin{aligned}& \int_0^t ds\int d\underline{x}\,\omega(x)\mathbf{E}_x\left|\int d\underline{y}\;\omega(\underline{y})\mathbf{E}_y\,\underline{k}\,(\underline{x}^{\nu}(s)-\underline{y}^{\nu}(s)) - \underline{k}(\underline{x}^{\nu}(s) - \underline{y}(s))\right| \\ & = \int_0^t ds\int d\underline{x}\,\omega_s^{\nu}(\underline{x})\int d\underline{y}\,\omega(\underline{y})\,|\,\mathbf{E}_y(\underline{k}(\underline{x} - \underline{y}^{\nu}(s)) - k(\underline{x}-\underline{y}(s)))| \\ & \le C\int_0^t ds\int d\underline{y}\,\omega(\underline{y})\mathbf{E}_y(\varphi(\underline{y}^{\nu}(s),\underline{y}(s))).\end{aligned} \tag{7.37}$$

Denoting

$$\gamma(t) = \int \omega(\underline{x}) \mathbf{E}_x(|\underline{x}^\nu(t) - \underline{x}(t)|) d\underline{x} \tag{7.38}$$

we have, after the use of Jensen inequality

$$\gamma(t) \leq \sqrt{2\nu} \|\omega_0\|_1 \sqrt{t} + C \int_0^t ds\, \tilde{\varphi}[\gamma(s)]. \tag{7.39}$$

Here, and above C denotes any constant depending only on ω_0.

Proceding as from (3.33) to (3.36) the proof is achieved.

For non positive ω_0 the proof goes along the same lines. □

Remark 3. In the proof of Theorem (7.2) we have implicitely assumed $\int |\omega_0(\underline{x})| |\underline{x}| d\underline{x} < \infty$ (this implies, with known estimates on displacements, that the l.h.s. of (7.36) makes sense). Nevertheless the finiteness of first moment is not necessary for we could use bounded distance, as in Section 3, to obtain the same result.

Notes

Some classical text books on stochastic process are [Gi.1], [Gi.2], [It.1], [Mc.1]. An elementary account on diffusion processes may be found in [Ro.1].

The study of a certain class of non linear parabolic equation by means of diffusion process techniques, has been proposed by Mc Kean [Mc.2].

A general review on the mathematical problems related to the Navier-Stokes equation is developped in the Temam's book [Te.1] where varios finite dimensional approximations of the solution are investigated (different from those proposed in the next Section). Theorem 4.1 presented in this Section follows the ideas of [Ma.1].

As regards to existence and uniqueness problem for Navier-Stokes eq.s see [Lad.1], [Te.1], [Sh.1].

For dimensions two the results are completely satisfactory even in presence of boundaries.

The situation is drastically different in dimensions three. The presence of the extra term $(\underline{\xi}\cdot\underline{\nabla})\underline{u}$ makes difficult to exclude a priori the arise of singularities. Thus existence and uniqueness theorems are available only in particular situations or for short times.

8. MEAN FIELD LIMIT AND PROPAGATION OF CHAOS FOR NAVIER-STOKES EQUATIONS

The arguments of the preceding section suggest to introduce a stochastic vortex dynamics to approximate the Navier-Stokes evolution in the same way in which the deterministic vortex dynamics simulated the Euler evolution. We discuss only the $D = \mathbb{R}^2$ case (the same arguments can be used to deal with the $D = T^2$ case) because the presence of boundaries (and the consequent production of vorticity) makes the analysis more difficult and, till now, only euristic results are available. The next section will be devoted to this important point.

Suppose, for notational simplicity, $\sigma = 1$ (and hence $\nu = \frac{1}{2}$) and the initial vorticity distribution ω satisfying $\omega \geq 0$ and $\int \omega \, d\underline{x} - 1$. Following the Euler case and taking into account the viscosity, we introduce the processes solutions of the following stochastic differential equation:

$$d\underline{x}_i^n(t) = \underline{u}_i^{\varepsilon}(X_n(t))dt + d\underline{b}_i(t) \tag{8.1}$$

where $i = 1 \ldots n$, $\underline{b}_i$ are n independent two dymensional brownian motions, $X_n(t) = \{\underline{x}_i^n(t)\}_{i=1}^n$, and

$$\underline{u}_i^{\varepsilon}(X_n(t)) = \frac{1}{n} \sum_{j=1}^{n} \underline{k}_{\varepsilon}(\underline{x}_i^n(t) - \underline{x}_j^n(t)) \tag{8.2}$$

where $\underline{k}_{\varepsilon}(\underline{x} - \underline{y}) \equiv \underline{k}_{\varepsilon}(\underline{x},\underline{y})$ has been defined in (3.4).

Let $X_n(t,X_n) = \{\underline{x}_i^n(t,X_n)\}_{i=1}^n$ be the 2n dimensional process satisfying (8.1) and starting almost surely from $X_n = \{\underline{x}_i^n\}_{i=1}^n \in \mathbb{R}^{2n}$ at time zero. The following measure (f bounded and measurable)

$$\mu_t^n(f) = \frac{1}{n} \sum_{i=1}^{n} \mathbf{E}(f(\underline{x}_i^n(t,X_n))) \tag{8.3}$$

is expected to approximate (in the weak topology sense) ω_t^ε, solution of the cutoffed Navier-Stokes equation with initial distribution ω, if

$$\mu^n(d\underline{x}) = \frac{1}{n} \sum_{i=1}^{n} \delta_{\underline{x}_i^n}(d\underline{x}) \tag{8.4}$$

approximates ω (in the same topology).

Established the problem, we realize that the situation is deeply different from the deterministic case. The reason is that, while μ_t^n defined in Section 3, *is* a weak solution of the cutoffed Euler equation, its analogue (8.3) *is not* a solution of the cutoffed Navier-Stokes equation in any form. In fact, by the use of the same arguments leading to (7.19)

$$\frac{d}{dt}\mu_t^n(f) = \frac{1}{n} \sum_{i=1}^{n} \mathbf{E}(\underline{u}_i^\varepsilon \cdot \underline{\nabla} f) + \frac{1}{2}\mu_t^n(\Delta f) \tag{8.5}$$

for all C^2, bounded functions. (8.5) has not the same structure of the Navier-Stokes equation because $u_i^\varepsilon(X_n(t,X_n)) \neq \int k_\varepsilon(\underline{x}_i^n(t) - \underline{y})\mu_t^n(d\underline{y})$. In other words the velocity field generated by a sample of the process is different from the expected (mean) field. Nevertheless one may hope (since we are scaling the field) to outline some indipendence property in such a way that in the limit $n \to \infty$ "tipical field" can be confused with mean values as if the processes $\underline{x}_i^n$'s would be independent. This feature, known as "propagation of chaos" in kinetic theory, holds in our context as asserted by the theorem below.

Theorem 8.1. Let

$$\tilde{\mu}_t^n(d\underline{x}) = \frac{1}{n} \sum_{i=1}^{n} \delta_{\underline{x}_i^n(t)}(d\underline{x}) \tag{8.6}$$

be a sequence of measure valued stochastic processes where $\{\underline{x}_i^n(t)\}_{i=1}^n \equiv \{\underline{x}_i^n(t,x_n)\}_{i=1}^n$ satisfies (8.1) with initial value $X_n = \{\underline{x}_i^n\}_{i=1}^n$.
If

$$\frac{1}{n} \sum_{i=1}^{n} \delta_{\underline{x}_i^n}(d\underline{x}) = \mu_o^n(d\underline{x}) \xrightarrow[n\to\infty]{} \omega_o \quad \text{weakly} \tag{8.7}$$

then for all $t > 0$

$$\tilde{\mu}_t^n(d\underline{x}) \xrightarrow[n\to\infty]{} \omega_t^\varepsilon \quad \text{weakly} \tag{8.8}$$

almost surely, where ω_t^ε is the (measure valued deterministic process) solution of the cutoffed Navier-Stokes equation.

Before giving the proof, let us interpret the above theorem. Consider an initial profile of vorticity ω_o and take a discrete version of it, namely μ_o^n. Consider n Brownian motions $\{\underline{b}_i\}_{i=1}^n$ and with these construct the process $X_n(t)$, and hence the random measure $\tilde{\mu}_t^n$. Theorem 8.1 asserts that when n increases, $\tilde{\mu}_t^n$ is going to become almost everywhere constant and approachs ω_t^ε. This allows to approximate ω_t^ε with a tipical sample of $\tilde{\mu}_t^n$ and hence by a tipical sample of the finite dimensional process $X_n(t)$. This is interesting, for instance, for numerical purposes : what one sees by simulating only a sample of the stochastic vortex motion is, for fixed t and sufficiently large n, a good approximation of the Navier-Stokes equation.

Proof of Theorem 8.1. First we discuss the case of non

interacting vortices i.e. $\underline{k}_\varepsilon = 0$. Let us define for an arbitrary and fixed $T > 0$ the space $\Omega = C([0,T];\mathbb{R}^2)$ equipped with the σ-algebra of the Borel sets w.r.t. the topology for which Ω is a metric space, with metric function given by

$$d(b,w) = \sup_{t\in[0,T]} d(\underline{b}(t),\underline{w}(t)) \qquad b,w \in \Omega, \tag{8.9}$$

where

$$d(\underline{b}(t),\underline{w}(t)) = \min(1,|\underline{b}(t) - \underline{w}(t)|). \tag{8.10}$$

We denote by $M(\Omega)$ and $M(\mathbb{R}^2)$ the space of the Borel probability measures on Ω and on $\mathbb{R}^2$ respectively, by $P_x(db)$ the conditional Wiener measure starting from x, and by $d\lambda \in M(\Omega)$ the measure:

$$d\lambda(b) = \int \omega_o(d\underline{x})P_{\underline{x}}(db). \tag{8.11}$$

Notice that if $F: \Omega \to \mathbb{R}^2$ is defined by

$$F(b) = f(\underline{b}(t)), \qquad t \in [0,T], \tag{8.12}$$

for some measurable $f: \mathbb{R}^2 \to \mathbb{R}^2$, then $d\lambda_t \in M(\mathbb{R}^2)$, defined as:

$$\int d\lambda_t(\underline{x})f(\underline{x}) = \int d\lambda(b)F(b), \tag{8.13}$$

is the distribution of $\underline{x} + \underline{b}(t)$, where $\underline{b}(t)$ is a standard brownian motion, and $\underline{x}$ is the random variable whose distribution is given by ω_o. Therefore

$$d\lambda_t(\underline{x}) = (e^{\frac{1}{2}\Delta t}\omega_o)d\underline{x}\,. \tag{8.14}$$

Let us now introduce the spaces

$$\Omega_\infty = \prod_{i=1}^{\infty} \Omega_i \, , \qquad \Omega_i = \Omega \tag{8.15}$$

and the measures

$$\mathbf{P} = \bigotimes_{i=1}^{\infty} P_i \, , \qquad P_i = P_o \tag{8.16}$$

where P_o is the conditional Wiener measure starting from zero.

Denote by $\nu^n \in M(\Omega)$ and by $\nu_t^n \in M(\mathbb{R}^2)$ the measures:

$$\nu^n(db) = \frac{1}{n} \sum_{i=1}^{n} \delta_{b_i}(db) \tag{8.17}$$

$\{b_i - x_i\}_{i=1}^{\infty} \in \Omega_\infty$,

$$\nu_t^n(d\underline{x}) = \frac{1}{n} \sum_{i=1}^{n} \delta_{\underline{b}_i(t)}(d\underline{x}) \qquad t \in [0,T] \, , \tag{8.18}$$

and assume

$$\underline{b}_i(0) = \underline{x}_i \tag{8.19}$$

according to (8.7). Thinking of $\underline{b}_i(t) - \underline{x}_i$ as a standard Brownian motion distributed via P_o, one can look at ν^n and ν_t^n as a sequence of stochastic measures realized on the probability space $(\Omega_\infty, \mathbf{P})$.

We have

Theorem 8.2. Suppose (8.7) holds.

Then, for $\mathbf{P}$-almost all $\{b_i - x_i\} \in \Omega_\infty$

$$\lim_{n\to\infty} R(\nu^n, \lambda) = 0 \tag{8.20}$$

where, as usual, R denotes the KR distance on Ω (see def. (3.9)) with respect to the metric function d. In particular for all bounded continuous $f: \mathbb{R}^2 \to \mathbb{R}$

$$\lim_{n\to\infty} \frac{1}{n} \sum_{i=1}^{n} f(\underline{b}_i(t)) = \omega_0(e^{\frac{1}{2}\Delta t} f) \tag{8.21}$$

almost surely.

The above theorem is noting else than an application of the strong law of large numbers. It is rather intuitive and its proof is straightforward. We give the proof in the Appendix, togheter with other probabilistic considerations.

Coming back to the our (weakly) interacting case, we introduce a useful map linking the interacting case to the free one, reconducting the proof of Theorem 8.1 to that of Theorem 8.2.

For all $\rho \in M(\Omega)$ we define $x^\rho: \Omega \to \Omega$ by

$$\Omega \ni b \to \underline{x}_t^\rho(b) = \underline{b}(t) + \int_0^t ds \int d\rho(w) \underline{k}_\varepsilon(\underline{x}_s^\rho(b) - \underline{x}_s^\rho(w)) \tag{8.22}$$

The above mapping is well defined since, for all $b \in \Omega$, the integral equation (8.22) has a unique solution for the regularity of $\underline{k}_\varepsilon$.

Furthermore we define $\theta: M(\Omega) \to M(\Omega)$ by:

$$\int F(b)(\theta\rho)(db) = \int F(x^\rho(b))\rho(db) \tag{8.23}$$

Let us investigate two useful examples of the map θ.

<u>1th case</u> - Let λ be defined in (8.11), ω_t^ε the solution of the cutoffed N.S. eq.s and $\underline{x}_t$ the process solution of

$$d\underline{x}_t = \underline{x} + b(t) + \int_0^t ds \int d\omega_s^\varepsilon(\underline{y}) \underline{k}_\varepsilon(\underline{x}_t - \underline{y}) d\underline{y} \tag{8.24}$$

where $\underline{x}$ is distributed via ω_o. Then, following the results of Section 7, the pair $(\omega_t^\varepsilon, \underline{x}_t)$, when $\omega_t^\varepsilon(A) = \int \omega_o(dx)\mathbf{P}\{\underline{x}_t \in A | \underline{x}_o = x\}$, is unique. Hence it is easy to realize that $\theta\lambda = \omega^\varepsilon$, where ω^ε is the distribution of the process $x = \{\underline{x}_t\}_{t\in[0,T]}$, i.e.

$$\omega^\varepsilon(A) = \mathbf{P}\ (x \in A) \tag{8.25}$$

<u>2nd case</u> - Let ν^n be defined in (8.17). By direct computation $\tilde{\mu}^n = \theta\nu^n$ where $\tilde{\mu}^n(db) = \frac{1}{n}\sum_{i=1}^{n} \delta_{\underline{x}_i^n}(db)$ and $\{\underline{x}_i^n\}_{i=1}^n$, $\underline{x}_i^n = \{\underline{x}_i^n(t)\}_{t\in[0,T]}$ are defined in (8.6).

With this in mind Theorem 8.1 follows by continuity property of the map θ and by Theorem 8.2.

In general we have, for any $\mu,\nu \in M(\Omega)$:

$$R(\theta\mu,\theta\nu) \leq R(\mu,\nu)\exp\{L_\varepsilon(1+T)e^{L_\varepsilon T}\} \tag{8.26}$$

To prove (8.26) we introduce $P \in C(\mu,\nu)$ a joint representation of μ and ν and from this we induce $P_\theta \in C(\theta\mu,\theta\nu)$ a joint representation of $\theta\mu$ and $\theta\nu$ in the usual way:

$$\int P_\theta(db,dw)F(b,w) = \int P(db,dw)F(x^\nu(b),x^\mu(w)) \tag{8.27}$$

It follows:

$$|\underline{x}_t^\nu(b)-\underline{x}_t^\mu(w)| \le |\underline{b}_t-\underline{w}_t|+\int_0^t ds\left|\int d\nu(1)\underline{k}_\varepsilon(\underline{x}_s^\nu(b)-\underline{x}_s^\nu(1))\right.$$

$$\left.-\int d\mu(1)\underline{k}_\varepsilon(\underline{x}_s^\nu(b)-\underline{x}_s^\mu(1)\right|$$

$$+\int_0^t ds\int d\mu(1)\,[\underline{k}_\varepsilon(\underline{x}_s^\nu(b)-\underline{x}_s^\mu(1))-\underline{k}_\varepsilon(\underline{x}_s^\mu(w)-\underline{x}_s^\mu(1))]$$

$$\le |\underline{b}_t-\underline{w}_t|+\int_0^t ds\left|\int P(db',dw')\,[\underline{k}_\varepsilon(\underline{x}_s^\nu(b)-\underline{x}_s^\nu(b'))\right.$$

$$\left.-\underline{k}_\varepsilon(\underline{x}_s^\nu(b)-\underline{x}_s^\mu(w'))]\right|$$

$$+L_\varepsilon\int_0^t ds\; d(\underline{x}_s^\nu(b),\underline{x}_s^\mu(w)) \qquad (8.28)$$

where L_ε is defined in (3.21).

Defining

$$y(t) = \int P(db,dw)\sup_{s\le t} d(\underline{x}_s^\nu(b),\underline{x}_s^\mu(w)) \qquad (8.29)$$

and

$$\gamma(t) = \sup_{s\le t} d(\underline{x}_t^\nu(b),\underline{x}_t(w)) \qquad (8.30)$$

we have, for $t \le \tau$,

$$d(\underline{x}_t^\nu(b),\underline{x}_t^\mu(w)) \le \{\sup_{s\le\tau}|\underline{b}_s-\underline{w}_s|+L_\varepsilon\int_0^t ds\, y(s)\}e^{L_\varepsilon T} \qquad (8.31)$$

and hence

$$\gamma(t) \le \{d(b,w)+L_\varepsilon\int_0^t ds\, y(s)\}e^{L_\varepsilon T} \qquad (8.32)$$

Integrating by P(db,dw):

$$y(t) \leq \left(\int P(db,dw)\, d(b,w) \right) e^{L_\varepsilon t} e^{e^{L_\varepsilon T} L_\varepsilon} \qquad (8.33)$$

yielding (8.26) after minimizing on P.

We now conclude the proof of Theorem 8.1. By the use of (8.26), replacing μ and ν by λ and ν^n respectively, and (8.20) in Theorem 8.2, we obtain

$$\lim_{n\to\infty} R(\omega^\varepsilon, \tilde{\mu}^n) = 0 \qquad (8.34)$$

and in particular (8.8). □

As we have seen, the stochastic vortex dynamics introduced in (8.1) is a good approximation of the cutoffed Navier-Stokes eq.s. The next step is to see haw "good" is this approximation. The large fluctuations theory investigates deviations from the behavior described by the strong law of large numbers.

Let us first see how this theory works for independent random variables.

Theorem 8.3. Let us define, for a measurable set $A \subset M(\Omega)$

$$\alpha_n(A) = \text{Prob}\{\nu_n \in A\} \qquad (8.35)$$

and

$$I^o_\lambda(\rho) = \begin{cases} \displaystyle\int_\Omega \ln \frac{d\rho}{d\lambda}\, d\rho & \text{if } \rho \prec \lambda \text{ and } \frac{d\rho}{d\lambda} \in L_1(\rho) \\ \infty & \text{otherwise,} \end{cases} \qquad (8.36)$$

$$I_\lambda(A) = \inf_{\rho\in A} I^o_\lambda(\rho)\,. \qquad (8.37)$$

Then:

$$-I_\lambda(\mathring{A}) \leq \liminf_n \frac{1}{n} \ln \alpha_n(A) \leq \limsup_n \frac{1}{n} \ln \alpha_n(A) \leq -I_\lambda(\bar{A}) \,. \quad (8.38)$$

(Here $\mathring{A}$ and $\bar{A}$ denote interior and closure of A).

The above theorem gives a sharp estimate of how ν_n deviates from λ. We further discuss the meaning of Theorem 8.3 in the Appendix collecting all the probabilistic content of this Section.

Now we establish an analog of Theorem 8.3 in the interacting case, by the use of the mapping θ.

First we state a preliminary definition. For $\nu \in M(\Omega)$ we denote by $\nu_t \in M(\mathbf{R}^2)$

$$\nu_t(B) = \nu(\{b | \underline{b}(t) \in B\}), \quad B \subset \mathbf{R}^2, \quad B \text{ measurable.} \quad (8.39)$$

Furthermore, introducing $\underline{\xi}_t$, the process solution of

$$\underline{\xi}_t = \underline{x} + \underline{b}(t) + \int_0^t ds \int \nu_t(d\underline{y}) k_\varepsilon(\underline{\xi}_s - \underline{y}), \quad (8.40)$$

where $\underline{x}$ is distributed by ω_0 and $\underline{b}$ is a standard Brownian motion, we denote by $\bar{\nu}$ the distribution of ξ i.e.

$$\bar{\nu}(A) = \mathbf{P}(\{\xi \in A\}), \quad A \subset \Omega, \quad A \text{ measurable.} \quad (8.41)$$

Obviously ω^ε is the fixed point of the map $\nu \to \bar{\nu}$.

<u>Theorem 8.4</u>. Defining, for any measurable set $A \subset \Omega$,

$$\beta_n(A) = \mathbf{P}\{\tilde{\mu}_n \in A\} \quad (8.42)$$

and

$$I_{\omega^\varepsilon}(\nu) = \begin{cases} \int \ln\frac{d\nu}{d\bar{\nu}}d\nu & \text{if } \nu \prec \bar{\nu} \text{ and } \frac{d\nu}{d\bar{\nu}} \in L_1(\nu) \\ \infty & \text{otherwise} \end{cases} \quad (8.43)$$

$$I_{\omega^\varepsilon}(A) = \inf_{\nu\in A} I_{\omega^\varepsilon}(\nu) \quad (8.44)$$

Then:

$$-I_{\omega^\varepsilon}(\mathring{A}) \leq \liminf_n \frac{1}{n} \ln \beta_n(A) \leq \limsup_n \frac{1}{n} \ln \beta_n(A) \leq -I_{\omega^\varepsilon}(\bar{A}) \quad (8.45)$$

Proof. We follow [Ta.1].
First we notice that, if $\nu = \theta\rho$, $\rho \in M(\Omega)$.

$$\int \bar{\nu}(db)F(b) = \int \lambda(db)F(x^\rho(b)) \quad (8.46)$$

and $\rho \prec \lambda$ iff $\nu \prec \bar{\nu}$.
Moreover if $g = \frac{d\nu}{d\bar{\nu}}$:

$$\int d\rho(b)F(x^\rho(b)) = \int d\lambda(b)g(x^\rho(b))F(x^\rho(b)) \quad (8.47)$$

yielding

$$\frac{d\rho}{d\lambda}(b) = g(x^\rho(b)) \quad (8.48)$$

and hence

$$I^0_\lambda(\rho) = I_{\omega^\varepsilon}(\nu). \quad (8.49)$$

Since

$$\alpha_n(A) = \beta_n(\theta A), \tag{8.50}$$

the this follows easily by the use of Theorem 8.3. □

Remark. The functional $I_{\omega^\varepsilon}(\nu)$ is an expression of the distance of ν from the fixed point ω^ε.

Finally we notice that a diagonal procedure $\varepsilon \to 0$, $n \to \infty$ can be used, as in Section 5, to obtain results for the uncut-offed Navier-Stokes equations.

Theorem 8.5. In the hypotheses of Theorem 8.1, for a suitable sequence $\varepsilon = \varepsilon(n)$:

$$\lim_{n\to\infty} \mu_t^n(f) = \omega_t(f)$$

where μ_t^n is defined in (8.3), (8.1), (8.2) and ω_t is the solution of Navier-Stokes equations with initial datum $\omega_0 \in L_1 \cap L_\infty$.

We do not give the proof of the above theorem that goes along the same lines of Theorem 5.1 making use of Theorem 8.1.

Notes

The results of this Section improve those presented in [Ma.1].

The proof of Theorem 8.1 is based on the properties of the mapping θ introduced by Tanaka [Ta.1] for the study of large fluctuations. Theorem 8.3 is taken by [Ta.1] where also a Central Limit Theorem is obtained.

Mc Kean [Mc.2] introduced the concept of propagation of chaos for such a class of parabolic equations, obtaining results similar to those presented in this Section.

Successively Tanaka and others obtained, along the same ideas, results in different contexts. See [Ta.2], [Ta.3], [Sz.1] for Boltzmann equation, [Ca.1] for Burgers equation and [Sz.2] for the same class of parabolic equations with reflecting boundary conditions.

A stochastic Vortex dynamics has been considered by Chorin [Ch.2] (also in presence of walls) to the end of simulating the solution of Navier-Stokes eq.s numerically.

The results of this Section prove a part of the Chorin's algorithm. A deeper discussion on this point will be made in the next Section.

Appendix

Proof of Theorem 8.2. First we suppose $\omega_o = \delta x_o$. Then $\lambda = P_{x_o}$ and

$$R(\lambda,\nu_t^n) \leq R(\lambda,\omega^n) + R(\omega^n,\nu^n) \tag{A.1}$$

where ω_t^n denotes the stochastic measure:

$$\omega_t^n(db) = \frac{1}{n}\sum_{i=1}^{n} \delta_{\underline{b}_i-\underline{x}_i+\underline{x}_o}(db) \tag{A.2}$$

Then, by (8.7)

$$R(\omega^n,\nu^n) \leq \frac{1}{n}\sum_{i=1}^{n} |\underline{x}_o - \underline{x}_i^n| \longrightarrow 0 \quad \text{as } n \to \infty \tag{A.3}$$

We now introduce the random variables defined in (Ω,P_o)

$$\chi_{t_1\ldots t_k}^{A_1\ldots A_k} = \chi(\{\underline{b}(t_i) \in A_i,\ i=1\ldots k\}) \tag{A.4}$$

$$|\underline{b}(t_i) - \underline{b}(t_k)|^4 \tag{A.5}$$

for t_i rational numbers and A_i rectangles with rational corners. The above family of random variables is countable and by the strong law of large numbers

$$\frac{1}{n}\sum_{i=1}^{n} \chi_{t_1\ldots t_k}^{A_1\ldots A_k}(\underline{b}_i) = \nu^n(\chi_{t_1\ldots t_k}^{A_1\ldots A_k}) \xrightarrow[n\to\infty]{} \mathbf{E}(\chi_{t_1\ldots t_k}^{A_1\ldots A_k}) \tag{A.6}$$

$$\frac{1}{n}\sum_{i=1}^{n} |\underline{b}_i(t_k)-\underline{b}_i(t_j)|^4 = \nu^n(|\underline{b}(t_k)-\underline{b}(t_j)|)$$

$$\longrightarrow \mathbf{E}(|\underline{b}(t_k)-\underline{b}(t_j)|^4) = \text{const}|t_k-t_j|^2 \tag{A.7}$$

for $\mathbf{P}$-almost all $\{\underline{b}_i - \underline{x}_i\}_{i=1}^{\infty} \in \Omega_\infty$. $\mathbf{E}$ denotes expectation w.r.t. P_o.

From (A.6), (A.7), Thm. 2 p. 514 and Remark 1 p. 513 of [Gi.1], we see that ν^n is relatively compact and by (A.6) has a unique limit point that is λ.

This proves 8.20 in case $\omega = \delta_{x_o}$ and the same is obviously true for ω convex combination of δ measures.

For general ω, one can approximate ω by convex combinations of δ-measures and use the weak continuity of ω w.r.t. ω_o (see definition of 8.11). □

Let us now discuss Theorem 8.3. This is consequence of a theory studying deviations from typical behavior of random variables on abstract spaces (for examples spaces of trajectories to which we are interested on). See [Az.1] for a general review on the argument.

To give an idea of the result described in Theorem 8.3 we discuss heuristically the simplest case of a sequence $\{x_n\}_{n=1}^{\infty}$ of equally distributed, independent, mean zero, real valued, random variables. Defining $X_n = \frac{1}{n}\sum_{i=1}^{n} x_i$ we know, by the strong law of large numbers that $X_n \to 0$ almost surely. To investigate deviations from the right behaviour we consider

$$\mathbf{P}(X_n \geq a) = Q_n, \quad a > 0 \tag{A.8}$$

A simple application of the exponential Chebychev estimates gives us

$$Q_n \leq \exp - n\{ta - \ln \mathbf{E}(e^{tx_1})\} . \tag{A.9}$$

Implying

$$Q_n \le e^{-I_a n} \qquad \text{(A.10)}$$

where, after optimizing on t, we define

$$I_a = \sup_t \{ta - \ln \mathbf{E}e^{tx_1}\} \qquad \text{(A.11)}$$

On the other hand, if there exists t* such that,

$$I_a = t^*a - \ln \mathbf{E}e^{t^*x_1} \qquad \text{(A.12)}$$

we get

$$a = \mu(x_1) \qquad \text{(A.13)}$$

where

$$\mu(f) = \frac{\mathbf{E}(e^{t^*\sum_i x_i} f)}{\mathbf{E}(e^{t^*x_1})^n} \,. \qquad \text{(A.14)}$$

Hence

$$\begin{aligned} Q_n &= \int d\mu\, \chi\, (\textstyle\sum x_i \ge na) e^{-t^*\sum x_i} \mathbf{E}(e^{t^*x_1})^n \\ &\ge \left(\int d\mu\, \chi\, (na \le \textstyle\sum_i x_i \le n(a+\varepsilon))\right) e^{-t^*n(a+\varepsilon)} \mathbf{E}(e^{t^*x_1})^n \end{aligned} \qquad \text{(A.15)}$$

By the strong law of large numbers, because of (A.13), the integral in the r.h.s. of (A.15) approaches a finite number. Hence

$$\liminf_n \frac{1}{n} \ln Q_n \ge I_a - \varepsilon \qquad \text{(A.16)}$$

The generalization of the above ideas to the case of independent random variables taking values in more complicated spaces, is non trivial. We address the reader to the references for a deeper understanding of the subject.

9. THE PROBLEM OF BOUNDARY CONDITIONS, CHORIN METHOD AND CONCLUDING REMARKS

In the last two Sections we have discussed how to deal with Navier-Stokes equations in two dimensions in terms of the evolution of vorticity. We have treated only the case in which the fluid is on all $\mathbb{R}^2$ and it is clear that the above considerations extend to domains without boundary. We want now discuss the difficulties arising in the more interesting case of a fluid in a domain with boundaries.

Let D be an open set in $\mathbb{R}^2$, with smooth boundary ∂D. Then the evolution equations for an incompressible viscous fluid in D are:

$$\begin{cases} \frac{\partial\omega}{\partial t}(\underline{x},t) + (\underline{u}\cdot\underline{\nabla})\omega(\underline{x},t) = \nu\Delta\omega(\underline{x},t) \\ \omega(\underline{x},t) = \text{curl } \underline{u}(\underline{x},t), \quad (\underline{\nabla}\cdot\underline{u})(\underline{x},t) = 0, \quad \underline{x}\in D, \ t \geq 0 \\ \underline{u} = 0 \quad \text{on} \quad \partial D \end{cases} \tag{9.1}$$

As we have seen, in treating the Euler case it was not necessary to specific the behavior of an element of vorticity on the boundary. The velocity field, defined as $\underline{u}(\underline{x},t) = \int (\underline{\nabla}^{\perp} g_D)(\underline{x},\underline{y})\omega(\underline{y},t)d\underline{y}$, is automatically tangent on the boundary thus the vortices in the interior never meet the boundary because of the uniqueness of the characteristics. When a viscosity is added the situation changes. From the point of view of differential equations one sees that the presence of the second derivatives requires an extra condition on the boundary (namely the tangent component of the velocity equal to zero for physical reasons). From the point of view of stochastic processes we need to describe the behavior of the vorticity on the boundary to make sense to a process hopefully connected with the evolution problem (9.1).

Unfortunately no simple boundary conditions exist on ω in order to get $\underline{u} = 0$ on ∂D. In fact, denoting by $\underline{n}$ and $\underline{\tau}$ the outward unit normal and the unit tangent respectively, the condition $\underline{u}\cdot\underline{\tau} = 0$ implies

$$\nu\underline{\tau}\cdot\Delta\underline{u} \equiv \nu\underline{\tau}\cdot\underline{\nabla}^{\perp}\omega = \underline{\tau}\cdot\nabla p \quad \text{on} \quad \partial D \tag{9.2}$$

Hence:

$$\frac{\partial\omega}{\partial n} = \frac{1}{\nu}\frac{\partial p}{\partial\tau} \tag{9.3}$$

To understand the role played by the pressure in the fluid dynamical equations, one may use the following theorem.

Theorem. Any vector field $\underline{v}$ on D can be decomposed as

$$\underline{v} = \underline{w} + \underline{\nabla}p \tag{9.4}$$

where $\underline{w}$ is a divergence free vector field such that $\underline{w}\cdot\underline{n} = 0$ on ∂D. Moreover $\int_D dx\, \underline{w}\cdot\underline{\nabla}p = 0$ and this implies that such decomposition is unique.

The above theorem can be applied to the Navier-Stokes equation for the velocity field. Let P be the projection operator on $L_2(D)$ on the divergence free vector fields, parallel to ∂D. Since $\partial_t\underline{u} = P\partial_t\underline{u}$, then $\underline{\nabla}p = (1-P)(-\underline{u}\cdot\underline{\nabla}\cdot\underline{u} + \nu\Delta\underline{u})$.

Coming back to the interpretation of (9.3) it seems hard

to describe the boundary conditions in terms of the behavior of the vortices near the wall. In fact the normal derivative of the vorticity, that is the rate of production of vorticity on the boundary, is related to the solution itself in a not very simple way.

This is not surprising. The condition $\underline{u} = 0$ on the boundary creates new vorticity because the velocity of the fluid slows down near the wall.

There is a method, due to Chorin, based on the idea of describing the production of vorticity of the walls to construct the solution. This method was invented for numerical purposes , and was applied to the study of a slightly viscous flow.

Let us describe the basic ideas.

Consider a fluid in a domain D, whose distribution of vorticity at time zero is ω and with velocity field $\underline{u}$ satisfying the right boundary condition.

If we evolve ω following the Navier-Stokes equation in all $\mathbb{R}^2$, the vorticity ω_ε, at time $t = \varepsilon$, generates a velocity field $\underline{u}_\varepsilon$ (computed with the free Green function) not satisfying the boundary conditions. They can be restored in the following way. First one introduce a potential flow $\underline{v}_\varepsilon$ satisfying:

$$\begin{cases} \text{curl } \underline{v}_\varepsilon = 0 \\ \underline{\nabla} \cdot \underline{v}_\varepsilon = 0 \\ \underline{v}_\varepsilon \cdot \underline{n} = -\underline{u}_\varepsilon \cdot \underline{n} \quad \text{on} \quad \partial D \end{cases} \tag{9.5}$$

Such a flow may be constructed by the introduction of a potential i.e. $\underline{v}_\varepsilon = \underline{\nabla}\varphi_\varepsilon$. From (9.5) we have the Neumann problem

$$\begin{cases} \Delta\varphi_\varepsilon = 0 \\ \dfrac{\partial\varphi_\varepsilon}{\partial n} = -\underline{u}_\varepsilon\cdot\underline{n} \end{cases} \tag{9.6}$$

whose unique solution can be computed in terms of the solution of the single layer integral equation.

The velocity field

$$\tilde{\underline{u}}_\varepsilon = \underline{u}_\varepsilon + \underline{v}_\varepsilon \tag{9.7}$$

has vorticity ω_ε and satisfies $\tilde{\underline{u}}_\varepsilon\cdot\underline{n} = 0$ on ∂D.

To impose also the condition $\underline{u}\cdot\underline{\tau} = 0$ we modify the field $\tilde{\underline{u}}_\varepsilon$ cutting it outside D. Defining

$$\begin{cases} \bar{\underline{u}}_\varepsilon(x) = \tilde{\underline{u}}_\varepsilon(x) & \text{if } x \in D \\ \bar{\underline{u}}_\varepsilon(x) = 0 & \text{if } x \in \mathbb{R}^2/D \end{cases} \tag{9.8}$$

the vorticity relative to $\bar{\underline{u}}_\varepsilon$ has a vortex sheet of intensity $\tilde{\underline{u}}_\varepsilon\cdot\underline{\tau}$ due to the discontinuity of $\bar{\underline{u}}_\varepsilon$. Namely

$$\bar{\omega}_\varepsilon(d\underline{x}) \equiv \text{rot}\; \bar{\underline{u}}_\varepsilon d\underline{x} = \omega_\varepsilon(\underline{x})\chi(\underline{x} \in D)d\underline{x} + \tilde{\omega}_\varepsilon(d\underline{x}) \tag{9.9}$$

where the measure $\tilde{\omega}_\varepsilon(d\underline{x})$, concentrated on ∂D, is defined as

$$\tilde{\omega}_\varepsilon(A) = \int_{A\cap\partial D} \tilde{\underline{u}}_\varepsilon\cdot d\underline{l} \tag{9.10}$$

for all open sets $A \subset \mathbb{R}^2$.

The above procedure may be iterated $\bar{\omega}_\varepsilon \to \bar{\omega}_{2\varepsilon}$ and so on and an approximate solution ω_t^ε may be defined as

$$\begin{cases} \omega_t^\varepsilon = \bar{\omega}_{k\varepsilon} \quad \text{if } t = k\varepsilon \text{ for some } k > 0 \quad k \text{ integer} \\ \omega_t^\varepsilon = \text{solution of Navier-Stokes equation in } \mathbb{R}^2 \text{ with initial} \\ \quad \text{condition } \bar{\omega}_{k\varepsilon},\ k = \text{Integer part } [\frac{t}{\varepsilon}] \text{ up to the time } t - k\varepsilon. \end{cases} \tag{9.11}$$

What is expected to be true is that the sequence ω_t^ε is weakly convergent for $\varepsilon \to 0$ to a weak solution of the Navier-Stokes equation with right boundary condition.

We remark that the approximate solution ω_t^ε is, at least, well defined because the Navier-Stokes initial value problem with datum (9.9), makes sense. (See comments in the Notes).

Beyond the convergence problems arising with the whole method, it has to be remarked that it reduces to a free evolution and a vorticity production on the boundary, both the two steps being computable by converging finite dimensional models. In fact the stochastic vortex dynamics (that is practically used by Chorin in its algorithm) has been proved to be convergent for the free flow in the last Section.

It has to be remarked that Chorin method (that is grid free) presents particular practical advantages when the Reynolds number R is very high. In this case the boundary layer is so small (of the order $1/\sqrt{R}$) to make impossible the construction of a grid for the finite difference methods.

A preliminary step in trying to prove the convergence of the Chorin method, would be the construction of the solutions of the Navier-Stokes equations following the physical idea of the vortex sheet generation on the boundary, with the aim to obtain the right boundary condition.

We illustrate a possible attempt in this direction by

discussing the halfplane case, to avoid geometrical complications.

Consider the Navier-Stokes equation in $\bar{D} = \{x_1,x_2 | x_2 \geq 0\}$ with an explicit forcing term F, to be determined by means of the solution ifself and the boundary conditions:

$$\begin{cases} \dfrac{\partial\omega_t}{\partial t} + (\underline{u}\cdot\nabla)\omega_t = \Delta\omega_t + F_t \qquad (\nu = 1) & (9.12) \\ \\ F_t(x_1,x_2) = f_t(x_1)\delta(x_2) \ . & (9.13) \end{cases}$$

Because of the condition (9.13), any solution of the problem (9.12) is also a solution of the usual Navier-Stokes equations in $D = \{x_1 x_2 \mid x_2 > 0\}$.

We assume $|\underline{u}(\underline{x})| \to 0$ as $|\underline{x}| \to \infty$, hence $\underline{u}$ is determined only by its vorticity.

To make sense to the evolution problem (9.12), (9.13) we have to specify two boundary conditions for the Laplace operator. The first, to reconstruct $\underline{u}$ by ω. We choose the Dirichlet b.c.:

$$\underline{u} = -\underline{\nabla}^{\perp}\Delta_D^{-1}\omega \ , \qquad \underline{u} = (u_1,u_2), \tag{9.14}$$

to have

$$u_2(x_1,0) = 0 \ . \tag{9.15}$$

The second to specify how the vorticity behaves at the boundary. For this we have a relative freedom and assume the Newmann (reflecting) b.c. for geometrical simplicity. This means that the equations (9.12), (9.13) may by understood in all $\mathbb{R}^2$ with the conditions

$$\omega(x_1,x_2) = \omega(x_1,-x_2)\,, \quad \underline{u} = -\nabla^{\perp}\Delta_D^{-1}\omega^+$$
$$u_1(x_1,x_2) = u_1(x_1,-x_2)\,, \quad u_2(x_1,x_2) = -u_2(x_1,-x_2). \tag{9.16}$$

where ω^+ is the restriction of ω in the superior halfplane.

Finally, f_t has to be determined to satisfy the missing b.c. for $\underline{u}$ i.e.

$$\frac{\partial}{\partial x_2}(\Delta_D^{-1}\omega_t^+)(x,0^+) = 0 \tag{9.17}$$

It is not hard to realize that condition (9.17) is equivalent to the more convenient condition.

$$\frac{\partial}{\partial x_1}(\Delta^{-1}\omega_t)(x,0) = 0 \tag{9.18}$$

In fact $\Delta_D^{-1}\omega_t^+$ is also a solution of Newmann problem if (9.17) is verified and hence coincide with $\Delta^{-1}\omega_t|_{x_2>0}$ that turns out to be the solution of the Dirichlet problem, implying (9.18). The complete equivalence is then obtained by reversing the argument.

((9.18) could have been obtained just from the beginning: assuming the Newmann b.c. for the evolution of ω and computing $\underline{u}$ also with the Newmann b.c., than $u_1(x_1,0)$ would be authomatically zero by symmetry, and hence (9.18) is equivalent to $u_2(x_1,0) = 0$).

Supposing $\underline{u}(x,t)$ known, one can insert ω_t in (9.18) and obtain an equation for f that can be uniquely solved.

We do not perform completely the analysis here, but limit ourself to see in some detail, how the situation is going on for the simple case of the Stokes evolution.

In this case, eq. (9.18) reduces to

$$h_t(x_1) + (Lf)_t(x_1) = 0 \tag{9.19}$$

where

$$(Lf)_t(x_1) = \frac{\partial}{\partial x_1}\Delta^{-1}\int_0^t (G_{t-s}F_s)(x_1,0)$$

$$h_t(x_1) = -\frac{\partial}{\partial x_1}\Delta^{-1}(G_t\omega_o)(x_1,0) \tag{9.20}$$

$$\omega_{t=0}(x_1,x_2) = \omega_o(x_1,x_2), \quad G_t = \exp \Delta t.$$

Eq. (9.19) follows by inserting in (9.18) the formal solu tion of the Stokes equation:

$$\omega_t(\underline{x}) = (G_t\omega_o)(\underline{x}) + \int_0^t (G_{t-s}F_s)(\underline{x})ds \tag{9.21}$$

Eq. (9.19) can be explicitely solved and this yields a solution of the Stokes equations in the halfplane case. We have:

$$\hat{f}_t(k) = 2i \operatorname{sgn} k[(k^2+\frac{\partial}{\partial t})\hat{h}_t(k)+\frac{|k|}{\sqrt{\pi}}\int_0^t \frac{ds}{\sqrt{t-s}} e^{-k^2(t-s)}(k^2+\frac{\partial}{\partial s})\hat{h}_s(k)]. \tag{9.22}$$

Here $\hat{.}$ denotes, as usual, the Fourier transform.

The explicit form of $f = L^{-1}h$ given by (9.22) allows to perform sharp estimates in the non linear case by means of perturbative techniques. The above approach has been investigated in [Be.2].

Another way to produce vorticity is to consider also the action of an external known force.

Let us consider the simplest case of $D = T^2$ and the evolution equation:

$$\frac{\partial\omega}{\partial t} + (\underline{u}\cdot\underline{\nabla})\omega = \nu\nabla\omega + G \qquad (9.23)$$

where $G = \text{curl}\,\underline{g}$ and $\underline{g}(\underline{x},t)$ is a known external force. One can study the Eq. (9.23) by the same methods in Section 7. The solution of (9.23) is given by the formal expression

$$\omega(\underline{x},t) = (V_{t,0}\omega)(\underline{x}) + \int_0^t ds\,(V_{t,s}G)(\underline{x},s) \qquad (9.24)$$

where $V_{t,s}$, the Navier-Stokes operator, is defined as

$$(V_{t,s}f)(\underline{x}) = \mathbb{E}(f(\underline{x}(t;\underline{x},s))) \qquad (9.25)$$

and $x(t;\underline{x},s)$ is the process, starting almost surely from $\underline{x}$ at time s and solving

$$d\underline{x}(t) = \underline{u}(x(t),t)dt + \sigma d\underline{b} \qquad (9.26)$$

with $\underline{u}(\underline{x},t) = \int \nabla^{\perp} g_{T^2}(\underline{x},\underline{y})\omega(\underline{y},t)d\underline{y}$.

Thus the solution $\omega(\underline{x},t)$ can be constructed by the study of a map, whose fixed point satisfies (9.23), with obvious modification of the methods discussed in Section 3 and 7.

Also algorithms based on the vortex model, which provide production of vortices to take into account the external force, can be made.

This model could be presumibly used to investigate asyptotic behavior in situations in which high wave numbers are relevant.

Notes

In a point of the Section we asserted the existence of the solution for the Navier-Stokes initial value problem relative to a vortex sheet. In case of positive viscosity the heat part is sufficient to guarantee existence and uniqueness of smooth solutions for all times.

As regards to Chorin method [Ch.2], see [Ch.3] for some related mathematical aspects.

The solutions of the Navier-Stokes equations on T^2, in presence of an external force have been numerically studied by truncating to five modes the Fourier transform of the equation [Bo.1]. The phenomenology of similar models has been widely investigated lately. The reasons of the interest in studying such evolution lies in trying to understand how to describe the apparence of turbulence in real fluids varying the Reynolds number ($\frac{\|g\|_\infty}{\nu}$ in this case). In contrast with Landau theory [La.1] describing the turbulence in terms of a high number of periods, Ruelle and Takens [R.1], [R.2], [Lan.1] connected turbulent hevarior with the apparence of invariant sets (called stronge attractors) in which the motion is chaotic and has strong ergodic properties.

These numerical experiments seem to confirm this hypothesis.

Obviously, this is not conclusive since the truncation to few modes is rough when R is large and in this region are just the interesting behaviors. See [Bo.2]. Nevertheless these models, as purely mathematical models, are interesting in itself. For considerations concerning the qualitative behavior of the full Navier-Stokes equation see [Ru.3].

Connected with this problematic is the research of invariant measures for the Navier-Stokes flow. The statistical theory of turbulence is very old and widely developped at level of formal manipulations (see [Ch.4] for a review and criticisms on these theories) but a theory of invariant measures is still missing.

Interesting results connecting statistical properties of the Navier-Stokes flow and fully developped turbulence have been established in [Fo.1] (see also references quoted there).

Invariant measures for the Euler flow in T^2 have been studied in [Al.1], [Bo.3] and [Bo.4].

These investigations are interesting also in connection with the turbulence theory if one believes that, once turbulence is developped, the Euler part is dominant. These measures generalize the Gibbs measure for vortex gas.

For general review of phenomenological aspects of two dimensional turbulence see [Kr.1].

Figure 1

Collapse of three vortices

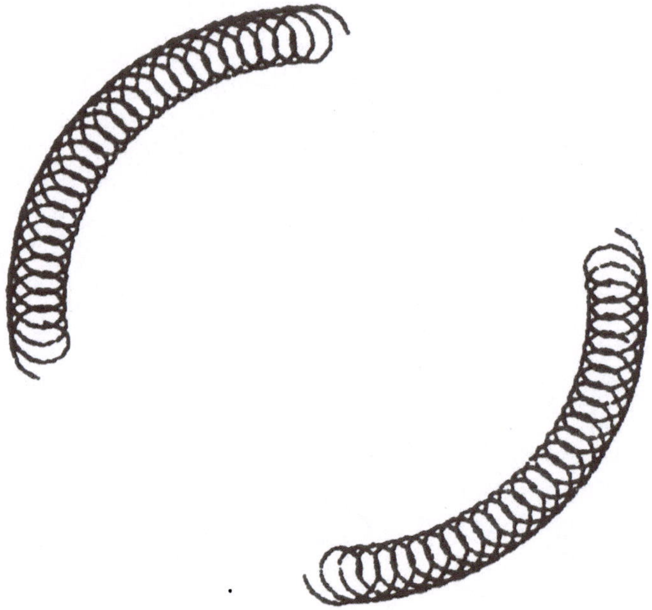

Figure 2

Two couples of weakly interacting vortices of the same intensity.

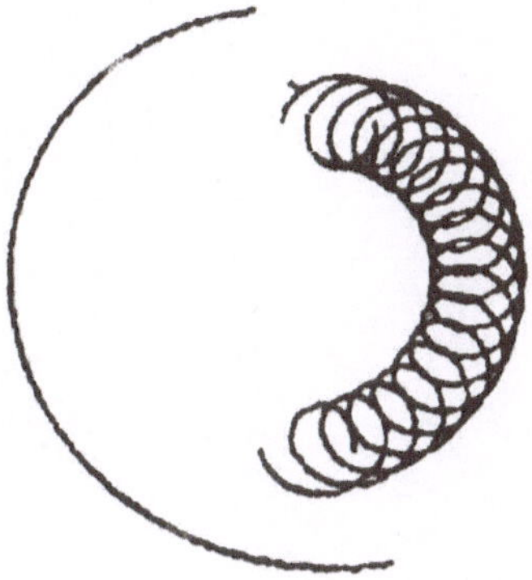

- Figure 3 -

A vortex in a circle weakly interacting with a pair. The three vortices have the same intensity.

References

[A.1] M. Aizenmann, Duke Math. J., 45 809 (1978).

[Al.1] S. Albeverio, M.R. de Faria, R. Høegh-Krohn, J. Stat. Phys., 20 p. 585 (1979).

[Ar.1] H. Aref, Physics of Fluid, 22 p. 393 (1979).

[Ar.2] H. Aref, N. Pomphrey, Physics Letters, 78 A p. 297 (1980).

[Ar.3] H. Aref, Ann. Rev. Fluid Mech., 15 (1983).

[Arn.1] V. Arnold, Methode Mathematique de la Mechanique Classique, M.I.R. Moscow (1976).

[Az.1] R.Azencott, Grandes Deviations et Applications, Lect. Notes in Math., 774 Springer (1980).

[Bar.1] C. Bardos, J. Math. An. Appl., 40 p. 769 (1972).

[Ba.1] K.G. Batchelor, An Introduction to Fluid Dynamics, Cambridge University Press (1967).

[Bea.1] J.T. Beale, A. Majda, Math. Comp., 39, 1 (1982); 39, 29 (1982).

[Be.1] G. Benfatto, M. Pulvirenti, A diffusion process associated to the Prandtl Equation, J. Funct. Anal. 52, 330 (1983).

[Be.2] G. Benfatto, M. Pulvirenti, Generation of vorticity near the boundary in two dimensional flows. Rome Preprint (1983).

[Bi.1] G. Birchoff, J. Fisher, Rend. Circ. Mat., Palermo 8 p. 77 (1959).

[Bo.1] C. Boldrighini, W. Franceschini, Commun. Math. Phys., 64 p. 159 (1979).

[Bo.2] C. Boldrighini, Introduzione alla Fluidodinamica, Quaderni del C.N.R., Roma (1979).

[Bo.3] C. Boldrighini, S. Frigio, Atti Sem. Mat. Fis., Univ. Modena XXVII p. 106 (1978).

[Bo.4] C. Boldrighini, S. Frigio, Commun. Math. Phys., 78 p. 303 (1980).

[Br.1] W. Braun, K. Hepp, Commun. Math. Phys., 56 p. 101 (1977).

[Ca.1] P. Calderoni, M. Pulvirenti, Propagation of Chaos for Burgers' Equation, Ann. Inst. H. Poincaré. A vol. XXXIX p. 85 (1983).

[Ch.1] A.J. Chorin, J.E. Marsden, A Mathematical Introduction to Fluid Mechanics, Springer, New York, Heidelberg (1979).

[Ch.2] A.J. Chorin, J. Fluid Mech., 57 p. 785 (1973).

[Ch.3] A.J. Chorin, T.J.R. Hughes, M. McCracken, J.E. Marsden, Commun. Pure and Appl. Math., vol. XXXI p. 205 (1978).

[Ch.4] A.J. Chorin, Lectures on Turbulence Theory, Publish or Perish, Boston (1975).

[Ch.5] A.J. Chorin, J. Comp. Phys., 27 p. 428 (1978); SIAM J. Sci. Stat. Comput., 1 p. 1 (1980).

[Co.1] R. Courant, D. Hilbert, Methods of Mathematical Physics, Interscience, New York (1953).

[De.1] A. De Masi, N. Ianiro, A. Pellegrinotti, E. Presutti, A Survey of the Hydrodynamical behavior of Many Particle Systems, Studies in Stat. Mech. Vol. 11 "Non equilibrium Process". J.L. Lebowitz, E.W. Montroll, North Holland (1984).

[Do.1] R.L. Dobrushin, Sov. J. Funct. Anal., 13 p. 115 (1979).

[Do.2] R.L. Dobrushin, Teor. Veroyatn. Priloz., 15 p. 469 (1970).

[Dü.1] D. Dürr, M. Pulvirenti, Commun. Math. Phys., 85 p. 265 (1982).

[Fi.1] P.C. Fife, Arch. for Rat. Mech. and An., 28 p. 184 (1968).

[Fo.1] C. Foias, R. Temann, Commun. Math. Phys., (1983).

[Fri.1] A. Friedman, Partial differential equation of parabolic type, Prentice Hall (1964).

[Fr.1] J. Fröhlich, D. Ruelle, Commun. Math. Phys., 87 p. 1 (1982).

[Gi.1] I.I. Gihmann, A.V. Shorohod, Introduction à la Theorie des Process Aleatoires, M.I.R. Moscow (1980).

[Gi.2] I.I. Gihmann, A.V. Shorohod, The Theory of Stochastic Process, Springer, Heidelberg, Vol. I (1974); Vol. II (1975); Vol. III (1976).

[Ha.1] O. Hald, V. Mauceri Del Prete, Math. Comp., 32, p. 791 (1978).

[Ha.2] O. Hald, SIAM J. Numer. Anal., 16, 726 (1979).

[He.1] H. Helmholtz, Phil. Mag., 33, p. 485 (1867).

[Hu.1] T.J.R. Hughes, J.E. Marsden, A Short Course in Fluid Mechanics, Publish or Perish, Berkeley (1976).

[I.1] K. Ito, H.P. McKean, Diffusion Processes and their Sample Paths, Springer, Heidelberg (1965).

[K.1] L.V. Kantorovich, G.S.H. Rubinstein, Dokl. Akad. Nauk SSSR, 115 p. 1058 (1957).

[Ka.1] T. Kato, Arch. Rat. Mech. and An., 25 p. 95 (1967).

[KAM.1] A.N. Kolmogorov, Dokl. Akad. Nauk SSSR, 98 p. 527 (1954);
V.I. Arnold, Usp. Mat. Nauk, 18 p. 13 (1963);
J. Moser, Math. Ann., 169 p. 136 (1967).

[Ke.1] O.D. Kellog, Foundation of Potential Theory, Springer, Berlin, Heidelberg, New York (1967).

[Kh.1] K.M. Khanin, Physica, D 4 p. 261 (1982).

[Ki.1] G. Kirchhoff, Vorlesungen ueber Math. Phys. Teubener Leipzig (1883).

[Kr.1] R.H. Kraichnan, D. Montgomery, Rep. Prog. Phys., 43 p. 547 (1980).

[Lad.1] O.A. Ladyzhenskaja, The Mathematical Theory of Viscous Incompressible Flows, Gordon & Breach, New York (1969).

[La.1] L.D. Landau, E.M. Lifshitz, Fluid Mechanics, Pergamon (1968).

[Lan.1] O. Lanford III, Ann. Rev. Fluid Mechanics, 14 p. 347 (1982).

[Le.1] A. Leonard, J. Comp. Phys., 37 p. 289 (1980).

[Ma.1] C. Marchioro, M. Pulvirenti, Commun. Math. Phys., 84 p. 483 (1982).

[Ma.2] C. Marchioro, M. Pulvirenti, Euler Evolution for Singular Initial Data and Vortex Theory, Commun. Math. Phys. 91, 563 (1983).

[Ma.3] C. Marchioro, M. Pulvirenti, On the Singularities of the Newtonian two dimensional N-body problem. Rend. Acc. N. Lincei (1983). To appear.

[Ma.4] C. Marchioro, E. Omerti, Time Evolution of an Infinite Number of Vortices in a Strip. J. Stat. Phys. to appear (1983).

[Mc.1] H.P. Mc Kean, Stochastic Integrals, Academic Press (1969).

[Mc.2] H.P. Mc Kean, Lectures in differential Eq.s, Vol. II p. 177, A.K. Aziz Ed. Von Nostrand (1969).

[Me.1] R.E. Meyer, Introduction to Mathematical Fluid Dynamics, Wiley (1981).

[Ne.1] H. Neunzert, An Introduction to the Non-Linear Boltzmann-Vlasov Equation. Lectures given at Int. Summer School "Kinetic Theories and Boltzmann Eq." CIME Montecatini - Italy (1981).

[No.1] E.A. Novikov, Zh. Eksp. Teor. Fis., 68 p. 1868 (1975).

[O.1] O.A.Oleinik, Dokl. Akad. Nauk SSSR, 150 p. 116 (1963).

[O.2] O.A.Oleinik, Doklady, 166 p. 727 (1966); Dokl. Akad. Nauk SSSR, 174 p. 775 (1967).

[On.1] L. Onsager, Suppl. Nuovo Cimento, 6 p. 279 (1949).

[Po.1] H. Poincaré, Théories des Turbillons, George Carré (1983).

[Ro.1] Y. Rozanov, Processus Aleatories, M.I.R. Moscow (1975).

[Ru.1] D. Ruelle, F. Takens, Commun. Math. Phys., 20 p. 167 (1971).

[Ru.2] D. Ruelle, Proceedings of the Int. School of Math. Phys., Camerino (1974).

[Ru.3] D. Ruelle, Commun. Math. Phys., 87 p. 287 (1982).

[Saa.1] D.G. Saari, N-body collisions and singularities, Lect. Notes in Pure and Appl. Math. 70 p. 187 (1981).

[Sa.1] P.G. Saffman, G.R. Baker, Ann. Rev. Fluid Mechanics 11, 95 (1979).

[Se.1] J. Serrin, Arch. for Rat. Mech. and An., 28 p. 217 (1968).

[Sh.1] M. Shimbrot, Lectures in Fluid Mechanics, Gordon and Breach (1973).

[St.1] E.M. Stein, Singular Integrals and Differentiability Properties of Functions, Princeton Univ. Press. Princeton N.J. (1970).

[Su.1] C. Sulem, P.L. Sulem, C. Bardos, U. Frisch, Commun. Math. Phys., 80 p. 485 (1981).

[Sz.1] A.S. Sznitman, Acad. Sci. Paris, 295 p. 363 (1982), Eq. de type Boltzmann spatialement homogénes (to appear).

[Sz.2] A.S. Sznitman, An example of non linear diff. proc. with normal reflecting boundary conditions,(to appear).

[Ta.1] H. Tanaka, Limit theorems for certain diffusion process with interaction, (to appear).

[Ta.2] H. Tanaka, Stoch. diff. eq.s associated with the Boltzmann eq. Stochastic Anal., A Friedman, M. Pinski Ed., p. 301, Academic Press, N.Y., San Francisco, London (1978).

[Ta.3] H. Tanaka, Some probabilistic problems in the spatially homogeneus Boltzmann eq., Proc. of IFIP-ISI Conf., Bangalore (1982).

[Te.1] R. Temam, Navier-Stokes Equations, North Holland (1977).

[V.1] L.N. Vaserstein, Prob. Peredachi Inf., 5 p. 64 (1969).

[V.M.1] R. Von Mises, K.O. Friedrichs, Fluid Dynamics, Springer, Heidelberg (1971).

Lecture Notes in Physics

Vol. 173: Stochastic Processes in Quantum Theory and Statistical Physics. Proceedings, 1981. Edited by S. Albeverio, Ph. Combe, and M. Sirugue-Collin. VIII, 337 pages. 1982.

Vol. 174: A. Kadić, D.G.B. Edelen, A Gauge Theory of Dislocations and Disclinations. VII, 290 pages. 1983.

Vol. 175: Defect Complexes in Semiconductor Structures. Proceedings, 1982. Edited by J. Giber, F. Beleznay, J. C. Szép, and J. László. VI, 308 pages. 1983.

Vol. 176: Gauge Theory and Gravitation. Proceedings, 1982. Edited by K. Kikkawa, N. Nakanishi, and H. Nariai. X, 316 pages. 1983.

Vol. 177: Application of High Magnetic Fields in Semiconductor Physics. Proceedings, 1982. Edited by G. Landwehr. XII, 552 pages. 1983.

Vol. 178: Detectors in Heavy-Ion Reactions. Proceedings, 1982. Edited by W. von Oertzen. VIII, 258 pages. 1983.

Vol. 179: Dynamical Systems and Chaos. Proceedings, 1982. Edited by L. Garrido. XIV, 298 pages. 1983.

Vol. 180: Group Theoretical Methods in Physics. Proceedings, 1982. Edited by M. Serdaroğlu and E. Ínönü. XI, 569 pages. 1983.

Vol. 181: Gauge Theories of the Eighties. Proceedings, 1982. Edited by R. Raitio and J. Lindfors. V, 644 pages. 1983.

Vol. 182: Laser Physics. Proceedings, 1983. Edited by J.D. Harvey and D.F. Walls. V, 263 pages. 1983.

Vol. 183: J.D. Gunton, M. Droz, Introduction to the Theory of Metastable and Unstable States. VI, 140 pages. 1983.

Vol. 184: Stochastic Processes – Formalism and Applications. Proceedings, 1982. Edited by G.S. Agarwal and S. Dattagupta. VI, 324 pages. 1983.

Vol. 185: H.N. Shirer, R. Wells, Mathematical Structure of the Singularities at the Transitions between Steady States in Hydrodynamic Systems. XI, 276 pages. 1983.

Vol. 186: Critical Phenomena. Proceedings, 1982. Edited by F.J.W. Hahne. VII, 353 pages. 1983.

Vol. 187: Density Functional Theory. Edited by J. Keller and J.L. Gázquez. V, 301 pages. 1983.

Vol. 188: A.P. Balachandran, G. Marmo, B.-S. Skagerstam, A. Stern, Gauge Symmetries and Fibre Bundles. IV, 140 pages. 1983.

Vol. 189: Nonlinear Phenomena. Proceedings, 1982. Edited by K.B. Wolf. XII, 453 pages. 1983.

Vol. 190: K. Kraus, States, Effects, and Operations. Edited by A. Böhm, J.W. Dollard and W.H. Wootters. IX, 151 pages. 1983.

Vol. 191: Photon Photon Collisions. Proceedings, 1983. Edited by Ch. Berger. V, 417 pages. 1983.

Vol. 192: Heidelberg Colloquium on Spin Glasses. Proceedings, 1983. Edited by J.L. van Hemmen and I. Morgenstern. VII, 356 pages. 1983.

Vol. 193: Cool Stars, Stellar Systems, and the Sun. Proceedings, 1983. Edited by S.L. Balliunas and L. Hartmann. VII, 364 pages. 1984.

Vol. 194: P. Pascual, R. Tarrach, QCD: Renormalization for the Practitioner. V, 277 pages. 1984.

Vol. 195: Trends and Applications of Pure Mathematics to Mechanics. Proceedings, 1983. Edited by P.G. Ciarlet and M. Roseau. V, 422 pages. 1984.

Vol. 196: WOPPLOT 83. Parallel Processing: Logic, Organization and Technology. Proceedings, 1983. Edited by J. Becker and I. Eisele. V, 189 pages. 1984.

Vol. 197: Quarks and Nuclear Structure. Proceedings, 1983. Edited by K. Bleuler. VIII, 414 pages. 1984.

Vol. 198: Recent Progress in Many-Body Theories. Proceedings, 1983. Edited by H. Kümmel and M. L. Ristig. IX, 422 pages. 1984.

Vol. 199: Recent Developments in Nonequilibrium Thermodynamics. Proceedings, 1983. Edited by J. Casas-Vázquez, D. Jou and G. Lebon. XIII, 485 pages. 1984.

Vol. 200: H.D. Zeh, Die Physik der Zeitrichtung. V, 86 Seiten. 1984.

Vol. 201: Group Theoretical Methods in Physics. Proceedings, 1983. Edited by G. Denardo, G. Ghirardi and T. Weber. XXXVII, 518 pages. 1984.

Vol. 203: C. Marchioro, M. Pulvirenti, Vortex Methods in Two-Dimensional Fluid Dynamics. III, 137 pages. 1984.

Selected Issues from

Lecture Notes in Mathematics

Vol. 909: Numerical Analysis. Proceedings, 1981. Edited by J.P. Hennart. VII, 247 pages. 1982.

Vol. 912: Numerical Analysis. Proceedings, 1981. Edited by G. A. Watson. XIII, 245 pages. 1982.

Vol. 920: Séminaire de Probabilités XVI, 1980/81. Proceedings. Edité par J. Azéma et M. Yor. V, 622 pages. 1982.

Vol. 921: Séminaire de Probabilités XVI, 1980–81 Supplément: Géométrie Différentielle Stochastique. Proceedings. Edité par J. Azéma et M. Yor. III, 285 pages. 1982.

Vol. 922: B. Dacorogna, Weak Continuity and Weak Lower Semicontinuity of Non-Linear Functionals. V, 120 pages. 1982.

Vol. 923: Functional Analysis in Markov Processes. Proceedings, 1981. Edited by M. Fukushima. V, 307 pages. 1982.

Vol. 926: Geometric Techniques in Gauge Theories. Proceedings, 1981. Edited by R. Martini and E.M.de Jager. IX 219 pages. 1982.

Vol. 927: Y. Z. Flicker, The Trace Formula and Base Change for GL (3). XII, 204 pages. 1982.

Vol. 928: Probability Measures on Groups. Proceedings 1981. Edited by H. Heyer. X, 477 pages. 1982.

Vol. 929: Ecole d'Eté de Probabilités de Saint-Flour X – 1980. Proceedings, 1980. Edited by P.L. Hennequin. X, 313 pages. 1982.

Vol. 930: P. Berthelot, L. Breen, et W. Messing, Théorie de Dieudonné Cristalline II. XI, 261 pages. 1982.

Vol. 931: D.M. Arnold, Finite Rank Torsion Free Abelian Groups and Rings. VII, 191 pages. 1982.

Vol. 932: Analytic Theory of Continued Fractions. Proceedings, 1981. Edited by W.B. Jones, W.J. Thron, and H. Waadeland. VI, 240 pages. 1982.

Vol. 934: M. Sakai, Quadrature Domains. IV, 133 pages. 1982.

Vol. 935: R. Sot, Simple Morphisms in Algebraic Geometry. IV, 146 pages. 1982.

Vol. 936: S.M. Khaleelulla, Counterexamples in Topological Vector Spaces. XXI, 179 pages. 1982.

Vol. 937: E. Combet, Intégrales Exponentielles. VIII, 114 pages. 1982.

Vol. 938: Number Theory. Proceedings, 1981. Edited by K. Alladi. IX, 177 pages. 1982.

Vol. 942: Theory and Applications of Singular Perturbations. Proceedings, 1981. Edited by W. Eckhaus and E.M. de Jager. V, 363 pages. 1982.

Vol. 953: Iterative Solution of Nonlinear Systems of Equations. Proceedings, 1982. Edited by R. Ansorge, Th. Meis, and W. Törnig. VII, 202 pages. 1982.

Vol. 956: Group Actions and Vector Fields. Proceedings, 1981. Edited by J.B. Carrell. V, 144 pages. 1982.

Vol. 957: Differential Equations. Proceedings, 1981. Edited by D.G. de Figueiredo. VIII, 301 pages. 1982.

Vol. 963: R. Nottrot, Optimal Processes on Manifolds. VI, 124 pages. 1982.

Vol. 964: Ordinary and Partial Differential Equations. Proceedings, 1982. Edited by W.N. Everitt and B.D. Sleeman. XVIII, 726 pages. 1982.

Vol. 968: Numerical Integration of Differential Equations and Large Linear Systems. Proceedings, 1980. Edited by J. Hinze. VI, 412 pages. 1982.

Vol. 970: Twistor Geometry and Non-Linear Systems. Proceedings, 1980. Edited by H.-D. Doebner and T.D. Palev. V, 216 pages. 1982.

Vol. 972: Nonlinear Filtering and Stochastic Control. Proceedings, 1981. Edited by S.K. Mitter and A. Moro. VIII, 297 pages. 1983.

Vol. 978: J. Ławrynowicz, J. Krzyż, Quasiconformal Mappings in the Plane. VI, 177 pages. 1983.

Vol. 979: Mathematical Theories of Optimization. Proceedings, 1981. Edited by J.P. Cecconi and T. Zolezzi. V, 268 pages. 1983.

Vol. 982: Stability Problems for Stochastic Models. Proceedings, 1982. Edited by V. V. Kalashnikov and V. M. Zolotarev. XVII, 295 pages. 1983.

Vol. 989: A.B. Mingarelli, Volterra-Stieltjes Integral Equations and Generalized Ordinary Differential Expressions. XIV, 318 pages. 1983.

Vol. 994: J.-L. Journé, Calderón-Zygmund Operators, Pseudo-Differential Operators and the Cauchy Integral of Calderón. VI, 129 pages. 1983.

Vol. 999: C. Preston, Iterates of Maps on an Interval. VII, 205 pages. 1983.

Vol. 1000: H. Hopf, Differential Geometry in the Large, VII, 184 pages. 1983.

Vol. 1003: J. Schmets, Spaces of Vector-Valued Continuous Functions. VI, 117 pages. 1983.

Vol. 1005: Numerical Methods. Proceedings, 1982. Edited by V. Pereyra and A. Reinoza. V, 296 pages. 1983.

Vol. 1007: Geometric Dynamics. Proceedings, 1981. Edited by J. Palis Jr. IX, 827 pages. 1983.

Vol. 1015: Equations différentielles et systèmes de Pfaff dans le champ complexe – II. Seminar. Edited by R. Gérard et J.P. Ramis. V, 411 pages. 1983.

Vol. 1021: Probability Theory and Mathematical Statistics. Proceedings, 1982. Edited by K. Itô and J.V. Prokhorov. VIII, 747 pages. 1983.

Vol. 1031: Dynamics and Processes. Proceedings, 1981. Edited by Ph. Blanchard and L. Streit. IX, 213 pages. 1983.

Vol. 1032: Ordinary Differential Equations and Operators. Proceedings, 1982. Edited by W. N. Everitt and R. T. Lewis. XV, 521 pages. 1983.

Vol. 1035: The Mathematics and Physics of Disordered Media. Proceedings, 1983. Edited by B.D. Hughes and B.W. Ninham. VII, 432 pages. 1983.

Vol. 1037: Non-linear Partial Differential Operators and Quantization Procedures. Proceedings, 1981. Edited by S.I. Andersson and H.-D. Doebner. VII, 334 pages. 1983.

Vol. 1041: Lie Group Representations II. Proceedings 1982–1983. Edited by R. Herb, S. Kudla, R. Lipsman and J. Rosenberg. IX, 340 pages. 1984.

Vol. 1045: Differential Geometry. Proceedings, 1982. Edited by A.M. Naveira. VIII, 194 pages. 1984.

Vol. 1047: Fluid Dynamics. Seminar, 1982. Edited by H. Beirão da Veiga. VII, 193 pages. 1984.

Vol. 1048: Kinetic Theories and the Boltzmann Equation. Seminar, 1981. Edited by C. Cercignani. VII, 248 pages. 1984.

Vol. 1049: B. Iochum, Cônes autopolaires et algèbres de Jordan. VI, 247 pages. 1984.

Vol. 1054: V. Thomée, Galerkin Finite Element Methods for Parabolic Problems. VII, 237 pages. 1984.

Vol. 1055: Quantum Probability and Applications to the Quantum Theory of Irreversible Processes. Proceedings, 1982. Edited by L. Accardi, A. Frigerio and V. Gorini. VI, 411 pages. 1984.